T0338010

Water Resources Management

Water Resources Management: Principles, Methods, and Tools

Neil Grigg

For general information on our other products and services or for technical support, please contact our Customer Care Department within the United States at (800) 762-2974, outside the United States at (317) 572-3993 or fax (317) 572-4002.

Wiley also publishes its books in a variety of electronic formats. Some content that appears in print may not be available in electronic formats. For more information about Wiley products, visit our web site at www.wiley.com.

Library of Congress Cataloging-in-Publication Data applied for
Hardback ISBN: 9781119885962

Cover Design: Wiley
Cover Image: © zombiu26/Shutterstock

Set in 9.5/12.5pt STIXTwoText by Straive, Pondicherry, India

SKY10039241_120522

Contents

List of Figures *vii*
Preface *ix*

1   Water Resources Management   *1*

2   History of Water Resources Management   *9*

3   Water Infrastructure and Systems   *25*

4   Demands for Water and Water Infrastructure   *45*

5   Hydrologic Principles for Water Management   *67*

6   Water Balances as Tools for Management   *89*

7   Flood Studies: Hydrology, Hydraulics, and Damages   *103*

8   Water Quality, Public Health, and Environmental Integrity   *119*

9   Models and Data for Decision Making   *133*

10   Operations, Maintenance, and Asset Management   *149*

11   Water Governance and Institutions   *165*

12   Water Management Organizations   *173*

13   Planning Principles, Tools, and Applications   *189*

14 **Planning for Water Infrastructure** *203*

15 **Water Quality Planning and Management** *211*

16 **Planning for Sociopolitical Goals** *223*

17 **Environmental Planning and Assessment** *235*

18 **Economics of Water Resources Management** *241*

19 **Financing Water Systems and Programs** *261*

20 **Water Laws, Conflicts, Litigation, and Regulation** *289*

21 **Flooding, Stormwater, and Dam Safety: Risks and Laws** *309*

22 **Water Security: Natural and Human-Caused Hazards** *319*

23 **Integrated Water Resources Management** *331*

24 **Careers in Water Resources Management** *345*

Appendices
Appendix A
Units, Conversion Factors, and Water Properties *355*
Appendix B
Acronyms and Abbreviations *361*
Appendix C
Associations, Federal Agencies, and Other Stakeholders of the Water Industry *367*
Appendix D
Water Journals *373*
Appendix E
Glossary of Water Management Terms *377*

Index *395*

List of figures

1.1   Water resources management steps.   *3*
1.2   DPSIR depiction of a drought scenario.   *5*
3.1   Water infrastructure systems.   *26*
3.2   Dam configuration.   *28*
3.3   A simple reservoir guide curve.   *29*
3.4   Dam cross section with hydropower generation.   *29*
3.5   Urban water system.   *30*
3.6   Stormwater system.   *35*
4.1   Demands for water management along a regulated stream.   *47*
4.2   Distribution of water withdrawals and consumption in the United States.   *49*
5.1   Hydrologic processes.   *68*
5.2   A hydrologic time series showing weekly averaging.   *75*
5.3   Types of aquifers.   *77*
5.4   Rain gages in a triangular region.   *83*
5.5   Channel cross section and profile.   *85*
5.6   Pumping well by a stream.   *87*
6.1   Inputs and outputs for a water balance.   *90*
6.2   Water balance along a reservoir-stream system.   *93*
6.3   A farm irrigation layout.   *95*
6.4   Water balance used to study losses in a distribution system.   *96*
6.5   Sequence of water rights on a stream.   *99*
6.6   Stream-reservoir system.   *101*
7.1   Flood hydrograph parameters.   *106*
7.2   Triangular hydrograph.   *107*
7.3   Channel floodway.   *108*
7.4   Hydrographs of flood routing through a reservoir.   *110*
7.5   Inflows and outflows from a stormwater detention pond.   *110*
7.6   Inflow and outflow of detention pond.   *112*
7.7   Typical depth–damage curve for single family residential structure.   *114*
7.8   Converging watersheds.   *117*
7.9   Flood hydrograph for conversion to unit hydrograph.   *118*
8.1   SDG goals linked to One Health framework.   *121*
8.2   Water-related pathways to contamination and disease.   *124*

8.3   Community water supply showing access to sources.   *126*

8.4   Threats to water supply and pathways to consumers.   *127*

8.5   Definitions relating to water resources assessment.   *128*

9.1   Models for analysis and to comprise a decision support system.   *135*

9.2   Block diagram of a typical hydrologic model.   *136*

9.3   Information and knowledge hierarchy.   *138*

9.4   DSS acting as a digital twin.   *141*

9.5   Reservoir performance for (a) 50 TAF and (b) 150 TAF.   *145*

10.1   Maintenance and renewal in the facility lifecycle.   *150*

10.2   A basic water department organization.   *151*

10.3   SCADA operating environment.   *152*

10.4   Systems control and information interfaces in a smart water system.   *153*

10.5   Data-centered management of physical assets.   *156*

10.6   A basic five-step asset management process.   *157*

10.7   Data and information elements in asset management.   *158*

11.1   Institutional arrangements for water management.   *167*

11.2   Scopes of water governance and water management.   *168*

11.3   Complexity as a function of water management scale.   *169*

12.1   Map of water sector players.   *174*

12.2   Management functions of water sector players.   *175*

12.3   Coordination of federal agency roles in water management.   *177*

12.4   River basins of the United States with USACE and USBR areas shown.   *178*

13.1   General process of water resources planning.   *190*

13.2   Feedbacks and iterations in water planning process.   *193*

13.3   Levels and types of water resources planning.   *194*

14.1   Infrastructure life cycle.   *204*

15.1   DPSIR depiction of water quality changes and responses.   *213*

15.2   Water management influences on water quality.   *215*

15.3   Human and natural determinants of water quality.   *216*

15.4   How the Water Quality Act works.   *217*

16.1   Hierarchy of needs for water.   *225*

18.1   Scope of water resources economics.   *243*

18.2   Classification of public goods related to water.   *244*

18.3   Methods and tools of water resources economics.   *247*

18.4   Cash flow diagram showing annual payments equivalent to present value.   *249*

19.1   A water utility financial model.   *265*

19.2   Water use rate structures.   *271*

20.1   How laws affect water management.   *293*

20.2   Surface water rights allocation in a watershed.   *295*

20.3   How the SDWA works.   *298*

20.4   Drainage in pre- and post-development situations.   *308*

21.1   Different types of flooding and flood risk.   *310*

21.2   A DPSIR of flood risk.   *311*

22.1   Buildup of water risks.   *320*

22.2   Risk management in a water organization.   *325*

23.1   Water resources management process with descriptions.   *332*

Preface

Water resources management is a field of work where you can "do well while doing good." You can work on problems like water shortages, pollution, and flooding, and you can have an interesting, satisfying, and remunerative career. This work still excites me after more than 50 years, and this is an attempt to share what I've learned. The book is for students, professionals, and policy leaders involved with water in public agencies, consulting firms, law firms, and public interest groups.

Water problems require you to stretch your imagination beyond a narrow disciplinary focus. They are complex, but the starting point is at the practical level where water resources managers work. This requires application of solid principles and methods, along with awareness of real-world situations.

Today, you can find practically any information you might need on the Internet, but you must know what to look for and how to apply it. Usually, this involves searching with key words and phrases, rather than to hunt for references which change continuously. In that sense, the chapters list some key references, which are usually government sites that are not expected to change, but it does not include long lists of sources like in a research article. Not every statistic is referenced, but they all have definite sources. By the same token, many free technical resources are available on the Internet. Their addresses may change so they are usually not listed, but you can normally find the current sites with searches.

In teaching water resources management for several decades, it has been my experience that awareness of facts is important, but the way that students should demonstrate understanding is in answers to questions and solutions to problems. This leads to a dual approach to teaching water resources management, one part about facts and principles and the other with questions, problems, and case studies. This led to the design of the book with explanatory material and comprehensive sets of definitions, questions, and problem-solving methods. These apply across disciplines, and the book also includes a set of quantitative problems that are distributed among the chapters. Most questions and problems are simple because the intent is to illustrate basic concepts and computations for water resources management and not to explain details of water resources engineering or scientific hydrology. Not every question is answered in the text, but assignments to look up answers can be given. Sometimes this might involve research, and the question provides an entry portal to explore the concept.

The questions, problems, and case studies comprise academic material, and they are grounded in real world situations where I have personal experience. These include consulting, conflict management, policy development, and management of public water management agencies. They also include material provided by other experienced water managers and engineers.

The sequence of the book begins with an explanation of the elements and inputs to water resources management decisions and actions. This includes presentations of water demands, hydrology, and the links with environment and health. This is followed by a discussion of how water infrastructure systems are operated and managed. Then, the institutional arrangements for water management are explained, to include law, economics, and finance. These provide the underpinnings of water resources planning as a comprehensive approach to addressing problems and issues. Finally, ways to create integrative solutions to involving water, society, and nature are presented.

An appendix is provided for units and conversion factors. Both English and SI units are used on the chapters, but not all are converted because it would make for tedious reading. Also, an appendix is provided for acronyms, which are defined in the chapters as well. A glossary is provided to define the many phrases and words that are common in water resources management practice.

The book is dedicated to Victor Koelzer, who was an accomplished engineer and water resources planner. Vic created the graduate course in water resources planning and management at Colorado State University and mentored me to take it over. Now, more than 500 students have completed my course, and many have gone on to very successful careers. Vic focused on infrastructure planning, and the course has evolved over the years to include broader topics that align with the current water management environment with many social, legal, and financial challenges.

Neil S. Grigg
19 October 2022

1

Water Resources Management

Water Resources Management: A Vital and Interdisciplinary Discipline

The stewardship of water is, without doubt, one of our most important responsibilities because life would simply not exist without water supplies for the environment and society. Water can also be destructive when flood waters ravage rivers and coastal zones, while at the same time shaping the Earth and renewing natural systems. These aspects of water as an essential resource and a powerful natural force create the problem space where water resources managers work.

Water resources management has the core purpose of managing water in all necessary ways to serve society and the environment. This requires water resources managers to apply scientific and management knowledge to make important decisions in the broad public interest. They address a diverse range of problems and can enter the field from different disciplinary backgrounds, such as engineering, law, hydrology, ecology, chemistry, finance, and others.

The problem space for water resources management includes situations with different types, levels, and complexities. Just as water must be shared, problems about managing it involve multiple people and must be addressed collectively. In studying water resources management, we consider problems in the abstract, like "global water problems," and in specific situations, like "find the best way to expand our city's water treatment plant." So, problem identification depends on the situation involving your needs or those of another person or group. After the unmet needs are identified, then a problem statement will specify what the problem or unmet need is, who is responsible, and why it is important to solve the problem or meet the need.

Water resources management involves solution of analytical problems, synthesis of solutions, and coordinating among diverse interests. The issues are compelling, and the bar is high. They were summed up by President John F. Kennedy, who said "Anyone who solves the problem of water deserves not one Nobel Prize,

Water Resources Management: Principles, Methods, and Tools, First Edition. Neil Grigg.
© 2023 John Wiley & Sons, Inc. Published 2023 by John Wiley & Sons, Inc.

but two – one for science and the other for peace." The quote captures the challenge: water resources management involves science and requires conflict resolution to help with peace.

This chapter defines water resources management and introduces problem identification and solution approaches. It explains why water resources management is important and how it works by applying knowledge from engineering and science, as well as diverse disciplines focused on governance, planning, law, and finance, among other topics. A shared understanding of these is needed, and the book seeks to explain and illustrate this knowledge in the form of principles, tools, methods, and common situations faced by water resources managers.

Defining Water Resources Management

While there is no consensus definition of water resources management, in a broad sense it is a process to allocate and control water resources and water infrastructure to achieve economic, social, and environmental goals. Allocating water resources means to divide up wet water among competing users and to distribute its resource benefits, like hydroelectric energy or water for navigation, among stakeholders. Controlling water resources also involves infrastructure, such as dams, as well as non-structural problem-solving, such as through regulatory programs.

The apparent reason for lack of consensus about the definition of water resources management is that practitioners have different perspectives that depend on their responsibilities, and academics have different disciplinary perspectives. Types of responsibilities might include supplying water, regulating wastewater, or controlling floods, for example. Levels of responsibilities might range from specific tasks like turning a valve to comprehensive government policy-setting about water. Academic perspectives include engineering, earth science, law, and economics, among others. For purposes here, water resources management is considered as a general concept that includes the full range of perspectives of both practitioners and academics. Examples are provided throughout the book to illustrate them.

With this broad range of perspectives and applications, the concept of water resources management might seem too general. However, the connectivity of water issues among many situations demands such a general approach to serve collective needs. The need to address connectivity drives a continuing search for an integrative way to apply water resources management. Several names are used for this integrative approach, especially Integrated Water Resources Management (IWRM), which builds on basic steps to provide tools to plan systems and solve problems. IWRM is explained in more detail in Chapter 23.

It is useful to view water resources management as occurring at three levels. The lowest or operator level involves basic operations of equipment like pumps and valves. The intermediate level extends to making decisions about water allocation, diversion, treatment, and other processes. At the highest level, it focuses more on stakeholder views, political issues, regulatory and legal constraints, and relationships with other sectors and may be called IWRM or by similar names. The three levels might be labelled water management, water resources management, and integrated water resources management, but consensus about such names is unlikely to occur.

The focus here is on a generic approach that can be adapted to any of the levels, but mainly to the intermediate and higher levels. The lower level involves more structured situations with fewer variables. Figure 1.1 shows a conceptual model of the basic steps of this generic approach of water resources management to different types of problem scenarios and actions.

In a water resources situation requiring management, some issues will require attention. Common situations recur in different contexts, such as supplying water, managing water quality, reducing flood risk, or resolving conflicts. Sometimes the problem at hand is clear, but other times it is bundled with other issues. For example, a food shortage may be exacerbated by water scarcity, which stems from poor water resources management. The managers working on the food shortage must work with water managers to address the problem jointly.

In the process, stakeholders must decide to initiate action and determine what to do. This can involve a planning phase that can take on different forms. This phase is important because it is where stakeholders work out conflicts, determine strategies, assemble resources, and develop plans of implementation. Once a course of action is determined, different interventions may be involved, like construction or modification of infrastructure, regulatory actions, operational changes, or development of policy, among other actions. These may involve different roles and stakeholders. Once a situation has been mostly resolved, a period of monitoring and assessment is used to determine if the water resources management mechanisms

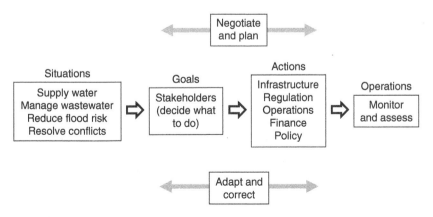

Figure 1.1 Water resources management steps.

are successful. If not, a corrective phase will be needed, and this can involve further activities from incremental adaptive changes to litigation and conflict resolution.

The actions shown in Figure 1.1 involve a workforce from diverse organizations with different disciplinary backgrounds and varying capacities. The organizations might include a water supply or wastewater authority, an irrigation district or company, a groundwater district, an electric power utility, a river basin authority, or regulator, for example. These involve different stakeholders and support groups, which also offer jobs in water resources management. As examples, consultants, government agencies, and utilities offer different kinds of jobs, and vendors of water management equipment also require skilled workers. Many kinds of water users, such as large farms and industries, also employ water managers.

Applications of Water Resources Management

Water resources management problems can often be understood from a picture of how the media reports them. For example, a story might focus on a drought that is allegedly caused by climate change. This situation can be depicted by use of a DPSIR diagram, which is a construct to show cause-effect relationships in social-ecological systems. The DPSIR displays the **Drivers** that affect water, the **Pressures** these drivers create, the resulting **State** of water resources, the **Impacts**, and the **Responses** from society.

As an example, Figure 1.2 depicts a basic DPSIR diagram for drought. Drivers like climate change and population growth create pressures, like depleted water storage and increased demands for irrigation water. These lead to declaration of a drought that causes crop failure, income losses, and food shortages. These urgent matters elicit responses such as mobilization of a drought response team, government relief, and new water storage projects.

Purposes of Water Resources Management

As problems like drought are confronted, solutions often have multiple purposes which show the range of applications of water resources management. Table 1.1 summarizes the most common water resources management purposes that occur. Each purpose has a name that will recur in discussions throughout the book.

As these purposes are pursued by water resources managers, a common set of scenarios or problem archetypes evolves. A way to organize these is as shown in Table 1.2.

In these scenarios, different water resources management responsibilities at the three levels mentioned earlier are evident. Combining the definition of water resources management with its purposes and scenarios leads to the set of management variables shown in Table 1.3.

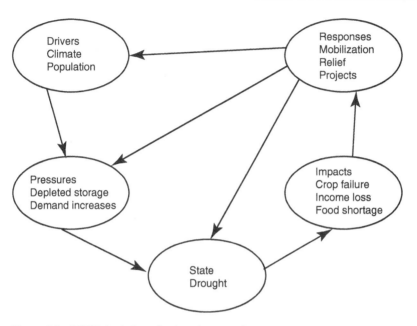

Figure 1.2 DPSIR depiction of a drought scenario.

Table 1.1 Purposes of water resources management.

Water service	Purpose
Water supply	Provide water for the economy, environment, and society.
Wastewater	Collect, treat, and dispose of wastewater.
Stormwater	Provide drainage systems and stormwater quality control.
Flood control	Reduce risk and protect against flood hazards.
Irrigation	Provide water for food production (livestock and aquaculture).
Environmental water quality	Regulate diversions and discharges to manage water quality.
Instream flows (E-flows)	Water to sustain natural ecosystems.
Hydropower	Manage instream flows for hydroelectric energy production.
Navigation	Control depth of water to support vessels on a waterway.
Recreation	Provide water in streams and lakes for recreation.

Table 1.2 Common scenarios of water resources management.

Scenario	Activities and functions
Policy development	Assess needs for programs, infrastructure, organizations, regulations, or reforms at different governance scales.
Watershed and river basin management	Align goals and plans of stakeholders with common interests in watersheds and river basins.
Infrastructure development	Plan and develop facilities by considering demands, alternatives, feasibilities, trade-offs, and impacts.
Operations	Analyze scenarios to plan for and operate water systems.
Program development	Develop management programs for utility finance, flood warning, water conservation, and other needs.
Regulation	Plan and implement regulatory controls for water quantity, quality, and land uses such as in flood control.
Conflict management	Mitigate conflicts due to transboundary flows, interbasin transfers, watershed issues, and organizational stovepipes.

Table 1.3 Management variables of water resources management.

Variable	How it characterizes water resources management?
Purposes	Explains water systems according to their purposes.
Methods	Methods such as construction and operation of infrastructure, use management systems, and apply functions such as planning and regulation.
Stakeholders	Groups involved in or impacted by decisions, such as water users associated with sectors and geographic areas.
Authorities	Organizations with roles, responsibilities, and functions such as policy development, management, operation, and regulation.
Sectors	Water-using groups for health, food, energy, and environment, among other uses. May be represented by authorities such as Ministry of Environment.
Scales	Alignment with watersheds, cities, counties, river basins, states, nations, and governmental units. May involve levels, as operator or director.
Stages	Steps in life-cycle of projects and programs, as in policy, planning, organization, implementation, operations, and renewal.
Context	Settings such as low- or high-income countries, varying governance effectiveness, urban or rural, growing or shrinking areas, and others.
Perspective	Perspectives differ according to incentives and impacts, such as among disciplines and roles in water management situations.

The benefit of viewing management variables this way is that the categories focus our understanding of different scenarios. Each category of variables has multiple dimensions, such as at micro- to macro-level scales. With an abundance of literature and media reports about water situations, the public and even water resources managers can become confused with so many facts and degrees of freedom, especially when conflicts are involved. With the list of variables, each situation can be described to fit coherently into the universe of problems and help us to draw on the appropriate knowledge bases to test solutions.

Questions

1 Explain the apparent differences between the levels of water resources management as explained in the chapter.

2 Formulate an example of a "multipurpose" water project and list at least three purposes.

3 For this project, list five or more stakeholder categories and their types of water management organizations.

4 Select a water issue such as lack of safe water, failed pollution control, or flood damage and prepare a DPSIR model to explain it in a general sense.

5 List five types of water issues that occur frequently around the world and provide examples.

6 Formulate and describe a scenario for water policy development that might occur at the state government level in the United States.

7 Why is water resources management fundamentally different as the scales of application go from small to very large?

8 Describe the level of precision required of water data at different stages of managing a water issue.

2

History of Water Resources Management

Introduction

People often express amazement at the rich history and multifaceted nature of water resources management with its different strands with technical, political, legal, and other parts. Many talented and fascinating people have contributed to its practices and lessons along the way. The technologies, infrastructures, and social institutions that emerged continue to evolve and provide useful lessons for today's water managers. While it developed in tandem with its counterparts in other sectors like energy and communications, water resources management has many connections with society that are distinct, especially with the environment and health.

People began to harness water to meet their needs early after the dawn of civilization, and water infrastructure was built as civilization advanced to create more sophisticated problem-solving methods through social institutions based on collective actions like, for example, development of mutual irrigation companies. Technologies and management tools are still evolving to address emerging problems, but social institutions remain challenging, such as when incentives for cooperation are missing and social institutions pose the greatest challenge to water governance.

The history of water resources management is extensive, and the material presented here is selective and influenced by the experience of the writer. Despite this, the chapter can serve as an outline of main topics and the reader explore them further through Internet searches and other sources.

Water Use Sectors

Water management examples include supplies for cities, farms, energy, the environment, and other purposes. Water supply is the paramount purpose, but management of wastewater and stormwater are also very important in urban

Water Resources Management: Principles, Methods, and Tools, First Edition. Neil Grigg.
© 2023 John Wiley & Sons, Inc. Published 2023 by John Wiley & Sons, Inc.

water systems. Water-related public health depends as well on plumbing systems in buildings. Big dams were constructed, and groundwater sources exploited to add to the capacity to store and provide water supplies, and many of the dams also provide flood control services. Programs for water quality and environmental protection depend on the regulatory environment and water laws. The management structure of water utilities and government agencies evolved to embrace these many needs and the private sector supports the public sector in solving water issues.

Development of water supply systems began with primitive facilities to meet the needs of early tribes and villages. Early people often settled near water sources, while powerful nobles and royalty developed supply systems for their castles and fortresses. Ancient settlements had water supply systems, such as the water tunnel for Jerusalem or the Qanats that provided water to Persia by tapping aquifers and transmitting the water through tunnels by gravity over long distances. By its peak, Rome had developed its aqueduct system to operate by gravity and its cities had sophisticated water systems. Some, like in Pompeii, had distribution systems that resemble ours today in many ways.

In Europe, older cities like London and Paris have long histories of water supply development that mirror the growth of the urban areas. Typically, private companies developed systems based on wells, but London began to develop more extensive supply systems by the early 1600s. In Paris, early residents took water from the River Seine, and during the eighteenth century an organized water system was under development. Technologies for water supply were evolving, including small dams, tunnels, wood pipes, and cast iron pipes that date back to the 1600s using technologies developed for making cannons.

In the United States, development of urban water supply systems in East Coast cities mirrored those in Europe, although with later starting dates. The history of many of them be found with Internet searches. For example, Boston's system is generally credited as the first major water works, having evolved in the eighteenth century. Philadelphia initiated a water supply system in 1798 after a yellow fever epidemic. It had pumping facilities driven by horses. New York's water system was developed shortly after 1800 through a venture that involved a banking scheme. Then, during the nineteenth century, more cities developed water supply systems. For example, Denver's system began soon after the 1859 gold rush with major facilities being developed as early as the 1880s.

Early water supply systems were not very safe, and this continued until development of water treatment systems that stem from public health discoveries based on microbiology and epidemiology of water-borne diseases. Water supply utilities became more sophisticated as municipal systems expanded. Also, many private water companies were organized and became important players in the water industry. For example, the corporate conglomerate American Water was created by mergers and acquisitions of private water companies. Such large private water companies operate as business conglomerates across the globe. Meanwhile, water services remain spotty among nations, and billions of people globally lack the levels of service enjoyed by people in high income countries.

Now, the focus is on developing complex water systems for sprawling cities and to extend universal safe water service to the world's population of about eight billion people. Urban systems have massive infrastructures for water storage, treatment, and distribution, but most water supply systems are outside of large cities and range from individual wells to small community systems. The levels of water safety, reliability, and affordability vary considerably, and large numbers of people remain unserved by organized water systems.

Wastewater systems developed along different trajectories than those for water supply. Prior to the 1880s, in-house toilets could not be connected directly to sewers because odors would waft back into buildings. After development of the P-trap seal, it became possible to locate bathrooms inside of homes and to send wastewater effluents to storm drains. By the late nineteenth century, the in-house plumbing systems we know today had begun to emerge. When wastewater was discharged, it quickly overloaded cesspools and privy vaults. This led to development of combined sewers as a solution to carry away domestic waste as well as drainage water.

During the early part of the twentieth century, most U.S. communities discharged wastewater through their sewer collection systems to nearby streams without treatment. Beginning in the 1950s, construction of wastewater treatment plants increased, and with the 1972 Clean Water Act their development accelerated. Impetus for this Act was driven by public impatience with visible pollution in the nation's waterways and rising concern about public health. Most cities now operate their wastewater systems as utilities and their performance is regulated by state agencies.

With more interest in sustainable development, there is a trend toward integrating wastewater systems with other waste streams, mainly food, to create holistic approaches to resource management. Water reuse systems have also advanced, both to extend water supplies and to reduce pollution in receiving waters.

Stormwater systems evolved from simple drainage facilities to the sophisticated multipurpose systems of today. Until the end of World War II, cities in the United States had grown slowly from their urban cores, and there was little attention to stormwater systems other than to provide limited drainage. With rapid urban development of the 1950s, stormwater systems began to attract notice. During the 1960s, the technical field of urban hydrology began to develop and some cities developed "blue-green" systems with open space along with closed conduits. The 1968 Flood Insurance Act introduced floodplain management to the stormwater mix. After the Clean Water Act in 1972, stormwater quality became a central issue to add to drainage and floodplain management. As a result of tax revolts, cities began to organize stormwater utilities in the 1980s. Current management of stormwater systems focuses on meeting multiple purposes including operation of combined sewer systems.

The earliest irrigation systems date back nearly 6000 years in Asia and the Middle East, such as in the Indus Valley and in Egypt along the Nile Valley. In many parts of the world, irrigation and farming have changed little in recent centuries and some ancient systems are still in operation. The older systems featured simple diversion structures, primitive devices to lift water, canals, and gravity irrigation distribution systems.

As systems developed, farmers banded together in water user and mutual associations to build ditches and shared facilities. Today, giant districts have been organized to provide irrigation water. Some of these districts also provide electric power and operate with large budgets and responsibilities. An example is the Imperial Irrigation District in California, which is the largest district in the United States and irrigates some 500 000 acres (202 000 ha) of land.

Today's systems have moved toward more automation and efficiency with "smart" irrigation systems. While gravity systems are still use, sprinklers and micro-systems apply water more efficiently. In smart systems, microprocessors control the flow of precise quantities of irrigation water.

Water-power has been utilized by people for thousands of years and ancient examples include water wheels to grind wheat more than 2000 years ago and water screws to lift irrigation water. Modern turbines were developed beginning in the mid-1700s to create more possibilities for harnessing the power of water. They were combined with generators beginning in the 1880s to enable businesses to provide street lighting and streetcar operation using direct current technology. Later, development of alternating current methods enabled the transmission of energy over long distances to initiate the electric power industry. By the early 1900s, hydropower furnished more than 40% of US electric power. A well-known example is at Niagara Falls, where power production began in 1895.

After World War I, electric power development focused on thermal plants, and hydropower declined as a percentage to less than 10% of US production now. However, hydropower is still important because it can be switched on and off quickly to provide peaking power.

The 1901 Federal Water Power Act was revised in 1930 to create the Federal Power Commission, which is now named the Federal Energy Regulatory Commission or FERC. The Tennessee Valley Authority was authorized in 1933, the same year that construction by the USBR began on Grand Coulee Dam, which has an installed capacity of 6.8 GW. In 1983, the Itaipú powerplant in Brazil and Paraguay became the largest plant in the world at 12.6 installed gigawatts. This has been supplanted now by China's Three Gorges Dam with 22.5 GW installed.

The Public Utility Regulatory Policies Act was passed in 1978 to encourage small-scale power production. In 1986, the Electric Consumers Protection Act amended the Federal Power Act to consider energy conservation, fish, wildlife, and recreation as well as power values when evaluating license applications. FERC has licensed many hydroelectric projects with dams and impacts on waterways.

Many power installations pre-date water regulation, and it is now harder to gain permission for hydro facilities and even to renew existing facilities may require litigation. With global warming looming over conventional fossil sources, the nation's energy future is uncertain. In any case, the interdependencies between water and energy will continue to be important drivers in water management decisions.

Inland navigation on rivers and dug canals has ancient roots, and in early civilizations, it could be easier to travel by water than over land. England had an impressive inland waterway system by the 1800s. In the United States, development of the Erie Canal was a pivotal event. Built between 1817 and 1825, the canal was successful. The Erie Canal company built the longest canal in the world, 364 miles (586 km), through largely unsettled wilderness. It had large economic impacts, both in opening trade with to the west and in making New York's harbor the most important in the nation.

The Suez Canal, developed in the 1860s as a private enterprise, was the first major constructed canal to link maritime navigation between seas. It has no locks because it connects the Mediterranean and Red Seas. It was followed by the Panama Canal, which opened in 1914. This canal was constructed under direction of the USACE and includes a lock system to lift vessels across the Isthmus of Panama. The makes the canal dependent on water management because floods and droughts can affect its water supplies.

The United States initiated its programs to manage inland navigation during the nineteenth century, with much of the emphasis on national defense as well as commerce. Today the United States has an extensive system of inland waterways, and most are maintained and operated by the USACE.

Dams

Early people understood the benefits of water storage, and began to build dams in the Middle East as early as 3000 BCE. As the Romans made advances in hydraulic engineering and dam construction technologies, they built gravity dams and the world's first arch dam in the southwest of France in the first century BCE. The Romans also built buttress dams on the Iberian Peninsula. One Roman gravity dam built about the first or second century AD still supplies water to Meriden, Spain. Asian people also developed dam engineering technologies, and by 400 BCE earth dams are built in what is Sri Lanka today. Dam construction resumed in Europe during the Middle Ages, Advances in geotechnical and structural dam engineering were made and concrete dam design technologies improved, especially for gravity dams.

The largest dams were built in the twentieth century with engineering advances like those evident in Hoover Dam on the Colorado River. It was originally named Boulder Dam and built as a public works project during the 1930s Great Depression. The gravity and arch structure is 726 feet (221 m) tall and impounds a reservoir of 28 million acre-feet (34.5 BCM). Its record height has been exceeded by the tallest dam in the world in Switzerland, the Grande Dixence Dam at 935 feet (285 m), and by California's Oroville Dam at 770 feet (235 m).

During the twentieth century, thousands of dams were built in the United States. Major projects include high dams such as Hoover Dam and Grand Coulee Dam. The Corps of Engineers (USACE) and the Bureau of Reclamation (USBR) are major dam builders, along with private power companies and the Tennessee Valley Authority. Many small dams were also built by local governments, irrigation districts, and private parties for diverse water management purposes.

As the environmental movement dawned, resistance to dam construction increased, and it became more difficult to build them. Dam building declined after the 1960s and now dams are considered to have mixed blessings. On the one hand, they provide economic benefits, but on the other hand, the negative environmental and social effects spur opposition.

Dam safety became a more visible issue after the 1976 failure of the Grand Teton Dam, which was a USBR project in Idaho. Today, removal of dams is being considered in many places. Although it is difficult to construct new dams in the United States, many are under construction in developing nations. The future of large dams in the United States is in question. On the one hand, there is pressure to remove some of them, but on the other hand a deep drought in the West has depleted water storage and called into question the capacity of dams like Hoover and Glen Canyon on the Colorado River to meet future water needs.

Groundwater

Development of groundwater by digging or drilling wells goes back to antiquity as humans learned to exploit underground water supplies. With the development of pumping systems, people could lift water more easily, and with the advent of diesel and electric motors, the modern groundwater development era began. Modern well systems use turbine pumps and can lift large volumes of water from deep below geological strata and make groundwater use feasible to irrigate large areas. In some cases, pumping such large quantities has created problems such as land subsidence and saltwater intrusion.

Public Health Engineering

One of the most important developments in water technology was the development of public health engineering, which is closely related to fields with the names sanitary engineering and environmental engineering. Early people knew about the link between water and infectious disease, but in Europe during the Dark Ages health deteriorated, and water quality links to health were only recognized again during the 1800s when the field of microbiology emerged.

When Dr. John Snow linked an 1854 cholera outbreak to a single source of water in London, the modern era of water quality began to emerge. Treatment systems started with filtration in 1887, with disinfection by chlorination dating back to 1909. The 1912 Public Health Act included controls on drinking water quality, but it was not very enforceable. By then, waterborne infectious diseases were on the decline, but increasing chemical problems became drivers for the Safe Drinking Water Act (SDWA) in 1974. It was amended in 1986 and 1996, and new regulations are issued periodically.

On a smaller scale, the plumbing craft evolved with the earliest pipes and fittings. The word plumbing derives from the Latin for lead, or *plumbum*. The earliest flush toilet has been identified from the palace of King Minos on the island of Crete, dating back to about 1700 BCE. Rome had advanced systems, but after it fell, western civilization declined during the Dark Ages. While royalty enjoyed more comfortable lives than peasants, even kings and queens died from typhoid and dysentery.

New plumbing devices were invented as civilization advanced. The earliest known flush toilet of modern times was by Sir John Harington, who installed one in a castle of Queen Elizabeth I about 1595. The earliest patent for a flush toilet was in 1775 to Alexander Cumming. Modern plumbing and public health emerged during the latter part of the nineteenth century as venting and drainage procedures helped to make indoor plumbing more acceptable. Plumbers of this era were metal workers who made their own fittings. After 1900 many advances occurred, leading to the modern industry we have today. Plumbing has continued to develop as a trade and a business. While today's tools and methods have improved from the past, plumbing still requires a lot of hard work in dark, cramped spaces.

Water Quality and Environmental Protection

Public health engineering is part of environmental engineering, which includes broader aspects of environmental management. An ecological thread has been added and today's environmental managers deal with fisheries, environmental impacts, and wildlife, was well as public health. The emergence of emphasis on environmental management began with the work of President Theodore Roosevelt and other conservation-minded leaders around 1900. Emphasis waxed and waned until the 1960s, where environmental conflicts and new disclosures played important roles in development of new policies. After Earth Day in 1970, a number of significant new laws were passed. Along with the National Environmental Policy Act (NEPA), the Clean Water Act (CWA) was a transformational development in this period. NEPA led to many situations where environmental impact statements were the defining decision documents on important proposed developments. The CWA led to similar situations, such as permits that triggered requirements for environmental impact statements, stream cleanups, estuary restorations, and other important initiatives.

Flood Control

The history of flooding and flood control began in ancient times with the biblical account of Noah's Ark as a starting point. Severe weather events continued to cause disasters and hardship for people living along waterways. Historical accounts of big floods can be found in many records, especially of big rivers like the Nile or the Yangtze in China. There were no large-scale organized flood control efforts until modern times, mainly because government programs had not been established.

In the United States, the major flood event that alerted the nation to the seriousness of the problem was the 1889 Johnstown, Pennsylvania, flood. It was caused by a dam failure, and loss and deaths were terrible. Coastal flooding in 1900 caused massive damages and death on Galveston Island in Texas. The 1927 Mississippi River flood is famous both for its hydrological features and the political changes it caused in the country. These and other floods led to the 1936 Flood Control Act, which enabled the use of benefit cost analysis to analyze projects. Another significant flood control act was in 1944 which authorized development of the Missouri River reservoir system.

By the 1960s, US policymakers recognize that dams for flood control did not solve the problems and the pendulum swung toward nonstructural solutions. The Flood Insurance Act of 1968 was the pivotal development that spurred nonstructural solutions based on flood insurance. Experience since then has identified problems needing resolution, but the flood insurance program remains the main US response to flood risk.

Government Involvement in Water Resources Management

The federal government was not much involved in water resources management during the nation's early history, but the stage for federal involvement was set when interstate compact provisions were written into the US Constitution. The

Lewis and Clark expedition in 1803–1805 explored the West and noted the need for irrigation, which was to attract federal involvement later. The USACE took on early roles in maintaining navigation routes and harbors for purposes of commerce and national defense. Other federal participation of water management was controversial, however, just as it was in other "internal improvements," which was a name for infrastructure in those periods.

Westward expansion and desire to settle the arid lands with irrigation fueled new activity. This led in 1902 to the Federal Reclamation Act and establishment of the Bureau of Reclamation to help bring water for development of the dry regions. Now, the United States had its two major Federal water mission agencies, the USACE and the USBR. Also, the USGS had been organized to provide data and studies about water resources and the Weather Bureau was established.

The Johnstown, Pennsylvania and Galveston disasters stimulated federal interest in flood control. In 1917 Congress began to develop the first Mississippi Flood Control Program. Activities were expanded after the disastrous 1927 Mississippi Valley flood and the 1936 Flood Control Act. During the 1930s, public works were expanded, and the Tennessee Valley Authority was created for electric power, flood control, conservation, and economic development. The Dust Bowl had created havoc in the Midwest, leading to organization of the US Soil Conservation Service, now named the Natural Resources Conservation Service (NRCS). It has been active in water conservation and construction of small watershed projects, and its engineers developed many useful formulas and technical guides. The Flood Control Act of 1944 authorized the Pick–Sloan plan for the Missouri River Basin, a joint effort of the USACE and USBR with projects for irrigation, power, flood control, and recreation.

After World War II many pressing national water issues became apparent. A Senate Select Committee on Water Resources was organized to study them and develop policy recommendations. In 1961 it recommended that the federal and state governments prepare comprehensive plans for major river basins, to include streamflow regulation, recreation and fish and wildlife. This led to the Water Resources Research Act of 1964 and the Water Resources Planning Act of 1965.

During the 1960s rising concern about environmental issues led to creation of the Environmental Protection Agency (USEPA) in 1969. This was a significant development that signaled new emphasis on environmental protection with a water focus, along with air quality and solid wastes.

State governments in the East initially had little involvement in water resources. However, in the West, state governments were active in water development. The most famous case is the California State Water Project, which was a state initiative and operates in tandem with the USBR's Central Valley Project.

Each state in the West has an interesting story of how its water was developed and is managed. In the East, state governments started to become involved in health issues, and this led naturally to environmental work as well. Today, state governments are more active in water programs, playing important roles in filling the gap between the federal government and local water providers.

Local governments became more involved with water management as urbanization increased. Each city has its own story of developing water and wastewater systems. Local water districts were formed to meet specific needs, and many were created to operate federal projects and to become management units as services and authorities grew. An example is the South Florida Water Management District (SFWMD), which was originally developed as an answer to hurricane-induced flooding. Congress created the Central and Southern Florida Flood Control Project in 1948 and the Florida Legislature created the Central and Southern Florida Flood Control District, which became today's SFWMD.

Water Law

Each country has a unique history of its water law. Roman law is often cited for initial concepts about how water should be allocated and managed. Asian cultures develop their own practices, especially in regard to irrigation. Water law in the United States began with state primacy because of constitutional arrangements for state and federal relationships. Other than to interpret English common law about disputes such as local drainage, there was little development of water law in the East. However, in the West development of water laws were necessary to allocate scarce water supplies.

The story of how California water law began in the mining camps and how methods were transferred to Colorado resulting in the appropriation doctrine is told later in the chapter on water law. Concepts of the water court, which has been adapted in Colorado, had roots in Spain during the Middle Ages.

A signature development in US water law was development of the 1922 Colorado River Compact. The legal basis for interstate compacts is embedded in the US Constitution, and the Colorado River Compact is the earliest and most famous example of its implementation.

During the twentieth century, federal law increased in scope, and after the 1970s, federal environmental law came to dominate many decisions. Each state has developed its own body of water law, with increasing emphasis on addressing problems of water quality and drought.

Management Structure in the Water Industry

The early history of the water industry focused on individual responsibility and actions by the rich and powerful to provide their own water services. As cities developed, they required organized water system and water utilities began to emerge. Many cities have stories of how their water systems became organized, such as how the Denver Water Department was formed upon completing purchase of the private Denver Union Water Company by 1920. Denver's growth was built on availability of an adequate water supply. The history of the water system began when the city was a mining camp and residents relied on private wells or streams. Several small companies were formed and were later consolidated. The Denver Union Water Company took over their assets. The municipal acquisition began in 1907 when Denver tried to buy the company, although it took until 1920 for the system to be converted to public ownership.

Prior to about 1900, most water services were private. Mistrust of the private sector led to pressure toward conversion to public sector ownership. The pendulum swung back toward privatization during the 1980s, led by England's initiative to privatize its water industry. This trend did not dominate, however, and many public systems remain publicly owned and operate alongside private ones.

Creation of irrigation districts began in the West prior to 1900. The oldest one is on the Pecos River at Carlsbad and dates to about 1880. The Imperial Irrigation District in California traces its roots to 1911. Other types of water districts have been organized, with California and Florida having major examples.

State regulatory agencies had early roots in public health and emerged as major players in water management. To a large extent, their development stemmed from federal funding made available through the Water Resources Planning Act, the Clean Water Act, and the Safe Drinking Water Act.

Industry played an important part in shaping management practices as trade associations and professional societies emerged in the late nineteenth century. The largest and most influential one is the American Water Works Association (AWWA), which began in 1881. It spun off a research foundation in 1967 and later a charity named Water for People. The Water Environment Federation (WEF) began in 1926 through an effort to create a Sewage Works Association. It also created a research foundation, which merged with the Water Research Foundation to create a joint effort between the AWWA and WEF spinoffs.

The United States does not have a national water authority to plan and develop water resources, but in some countries a national water authority provides overall management. This is the case, for example, in Latin American countries, such as Brazil and Peru, both of which have national water authorities. In the United States, the closest example is the California Department of Water Resources, which plans, manages, and operates major projects. However, it is still subject to the higher authority of federal law and policy.

Science, Engineering, and Technology

The technical foundation for water management stems from early scientific and industrial developments in physics, chemistry, and development of equipment for pumping systems and instrumentation, among other developments.

Principles of fluid mechanics, hydraulics, and hydrology evolved from about in 1770, when Chézy published an equation for open channel flow with Manning developing an equation for the resistance factor later. The Darcy–Weisbach equation for resistance in pipe flow dates to the middle nineteenth century. The unit hydrograph was developed during the 1930s, and other equations and concepts have been developed with many being incorporated in software.

Early development of techniques for water treatment were developed at the Lawrence Research Station in Massachusetts. These included practical methods for treating wastewater and also slow sand filtration. Work in this era included developments by scientists who became famous for research and consulting engineering at the same time.

Research on water resources systems analysis was initiated at Harvard University during the 1950s. Arthur Maass and Maynard Hufschmidt wrote a paper in 1959 titled "In Search of New Methods for River Basin Planning," which is sometimes cited as the beginning of systems analysis for water management. Now, many new tools have emerged, with more promise as computing and data systems improve.

The water industry uses modern technological tools like advanced infrastructure, computers, and instruments to detect contaminants. These evolved slowly over the centuries to include dams, pipes, pumps, valves, treatment processes, plumbing fixtures, and others. For example, beginning in the seventeenth century, use of cast iron pipe emerged for city water systems. Much of it is still in use, with rehabilitation or replacement being necessary when it fails. A number of new pipe technologies have been developed and there are new methods for trenchless technologies, such as microtunneling, pipe bursting, and cleaning/lining methods.

Business of Water

While much of the history of water management is about public organizations, the private sector has been instrumental as well. Private water companies played an important part in the development of early water supply utilities, and some have grown to be giant corporations operating around the world. Vendors of equipment, services, and supplies have major influence in the water industry and their diverse histories include development of pumps, pipe rehabilitation, and consulting, among other equipment and services.

From the beginning of water systems development, consulting engineers were influential in developing infrastructure and management methods. For example, Dewitt Clinton prepared an early report for the historic Croton Aqueduct in

New York City. In recent decades, consulting engineering firms have increased in number and influence, compared to public sector engineers.

The influence of the nongovernmental sector of the water business has grown with initiation of many interest groups ranging from promotion of business to advocacy of public interest topics such as health, environment, and justice. The numbers of these organizations continues to grow, as they partner with interest groups, businesses, and governments to explore new ways to improve water management and equity.

Questions

1 Water industry associations play an important part in development of professionalism and the workforce. What is the largest and most influential water supply association in the United States?

2 What was the original name of Hoover Dam and what was the economic situation in the United States when it was constructed? What has been the major issue involving its role during the period after about 2020?

3 A significant dam failure occurred in the United States in 1976. It had major influence on water resources policy. What dam was that and who owned it?

4 What was President Theodore Roosevelt's main environmental legacy?

5 What was the context of Earth Day in 1970 and how did it affect water policy?

6 What flood disaster in 1889 had a major influence on US flood policy? What was the main policy outcome?

7 What was the driving force behind the 1936 Flood Control Act and what was its main provision that has affected economic analysis in water planning?

8 What new flood related law introduced floodplain management as a main objective in water resources management?

9 When was the Tennessee Valley Authority created, what was the political context, and what were the major objectives and outcomes?

10 Which two federal departments are the main dam construction and operation agencies?

11 Why is it said that interstate water compacts work under authority of the US Constitution? What is their legal basis and how is one used for the Colorado River allocations?

12 What was the main reason for creation of the Bureau of Reclamation?

13 The California State Water Plan is the largest state government water project in the United States. What is the companion federal project that manages much of California's water?

14 What was the impetus that gave rise to the creation of the South Florida Water Management District?

15 What was the political context for the creation of the US Environmental Protection Agency?

16 What was the role of hydropower at the dawn of the electric age?

17 What is the largest US irrigation district and what is its story?

18 What were the water policy studies that led to the Water Resources Planning Act of 1964?

19 Explain the allocation of responsibilities among federal and state governments for water laws.

20 What were the historical precedents that led to the appropriation doctrine of water law in the American West?

21 What is the Erie Canal? What is its significance in early water development?

22 Who was Dr. John Snow? What was his contribution to water and health?

23 Historically speaking, why was river development the first water management focus of the federal government?

24 Name the key historical hydrology and hydraulics developments that support engineering practice.

25 What was the contribution of the Harvard Water Program to advancements in water resources management?

26 Explain how the early storm drains developed and how the practice of stormwater management has changed.

27 What happened during the urban development of the 1950s that created large impacts on stormwater systems?

28 Where are the oldest water supply systems in the United States found?

29 What was the main historical driver that led to the SDWA?

30 What are Qanats and why were they developed for water supplies in the Middle East?

31 How did a Roman aqueduct work?

32 Explain the historical evolution of water reuse systems and why they were developed.

33 What was the significance of the P-trap water seal on building plumbing systems and wastewater management?

34 About when did accelerated construction of wastewater treatment systems begin and what were the driving forces?

35 What were the earliest water treatment developments?

3

Water Infrastructure and Systems

Introduction

Although water resources management includes non-structural approaches, it relies heavily on infrastructure, from large dams to smaller components such as tanks and pump stations. Water managers plan, develop, operate, maintain, and regulate these infrastructure systems and components to capture, store, transport, and process water. Many decisions about them are made by governance authorities in city councils, water boards, government agencies, and other organizations.

This chapter describes the configurations, operational characteristics, and issues of the most important water infrastructure systems and components. These topics involve a broad knowledge base because they span across different types of systems, equipment, and devices. The questions at the end of the chapter serve as integrating mechanisms among these knowledge bases to highlight the main principles required for successful management of them. Numerical problems involving the different infrastructures are included in other chapters.

Water Infrastructure Overview

The term water infrastructure means dams to many people, but there are more types and viewing them as a system in a watershed (Figure 3.1) shows how they are connected to each other and to natural water systems. The major components begin with a dam and reservoir to control the water releases. These serve instream purposes for hydroelectric generation, fisheries, and navigation, and some of the water is diverted to serve off-stream needs such as urban water supply and irrigation. Levees are shown for flood control purposes, and there are intersections such as

Water Resources Management: Principles, Methods, and Tools, First Edition. Neil Grigg.
© 2023 John Wiley & Sons, Inc. Published 2023 by John Wiley & Sons, Inc.

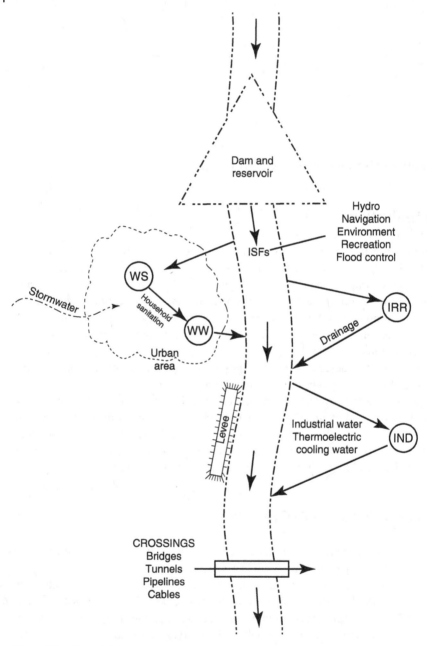

Figure 3.1 Water infrastructure systems.

pipelines and roads crossing the streams. Natural water infrastructure components, such as streams, aquifers, and wetlands, are important parts of the watershed system.

The major systems shown on the figure can be expanded to show subsystems for water supply, wastewater, and irrigation that include elements such as distribution systems with pumps, tanks, and valves, along with local hookups, service lines, and appurtenances. This expansion illustrates how water infrastructure systems are hierarchical, from major structures, like dams, down to small components, like valves in distribution systems.

The main variables that characterize water infrastructure systems are their functions and scales. The main functions are water conveyance, storage, treatment, control, and energy conversion (such as pumps and turbines). Scales range from individual sites to small community, large community, regional, or larger scales. For example, the function of water supply systems is to provide safe and reliable water to users, and this occurs at scales ranging from individual sites to megacities. Context is another important variable, and for water infrastructure systems it will change across situations such as major cities in high-income countries to small villages in low-income countries.

Dams, Reservoirs, and Hydropower Systems

Of the water infrastructures, dams are the largest structures and have the greatest impacts on natural systems. Despite these impacts, dams provide valuable water supplies where they may not be available otherwise. Dams impound water in storage reservoirs to serve multiple purposes, such as to carry water over from wet to dry periods or to provide head for hydropower generation in run-of-river reservoirs. Some dams provide hydroelectric power, which may be bundled with project purposes such as water supply and flood control. Many dams also serve navigation and have locks as part of their infrastructure.

Global estimates of the numbers of dams have been made but are not current. The World Register of Dams, initiated in 1958, serves as a reference for dams in different countries but is not complete. In the United States, the National Inventory of Dams, which is managed by the USACE, lists about 90 000 dams that are considered large enough to monitor. China has about the same number.

When describing a dam, you would specify variables such as height, construction method, and ownership. For example, a dam could be a low rockfill dry dam used for flood control and owned by a local government. The construction methods include earthfill or rockfill and concrete dams (gravity, buttress, and arch). Heights range from low to the world's highest, in Switzerland at 935 feet in height. Operational modes can be for storage, run-of-river, or dry dams, and purposes can range from water supply to flood control and more. Ownership can be by

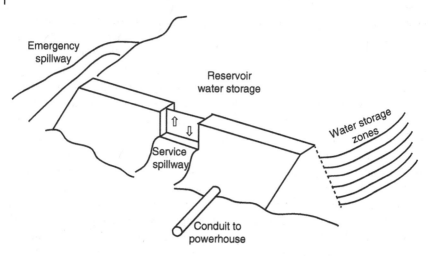

Figure 3.2 Dam configuration.

government or private parties. For safety purposes, dams can also be classified as high hazard, significant hazard, and low hazard, as in the US rating system.

The basic configuration of dams is shown in Figure 3.2. Most dams have a conduit for outlet works to release water for downstream purposes, including hydropower. Many dams have two spillways: one for operational use and the other for emergencies. The size of spillways is important because they must pass extreme flows or the dams might be overtopped and at risk of failure. The storage space in reservoirs may be allocated among purposes of water management such as water supply and flood control. For example, the series of reservoirs on the mainstem of the Missouri River has a permanent pool, a zone for multiple uses, a zone that may be for multiple uses or annual flood control, and a zone that is exclusively for flood control. Above that top zone, additional water may be stored as surcharge, but that will only occur during extreme emergencies and will be drained as soon as possible.

Reservoirs and lakes also have biologic zones, which affect water quality and aquatic life. These are defined by light penetration, proximity to shore, and nearness to the bottom. Water gets mixed in reservoirs because surface water can cool in winter, become heavy and sink, resulting in a lake turnover. During spring and summer, wind and waves create movement and the lake mixes different qualities of water and biological activity. These turnovers can create changes in lake ecology during the seasons.

Hydrologically, reservoirs are subject to evaporation and seepage. In hot and dry climates with long daylight hours, evaporation will be much larger than in cooler, humid climates with shorter daylight hours. Seepage from reservoirs depends on geology and can be significant when soils and underlying rock structures are porous.

Sedimentation occurs in reservoirs and reduces their capacity over time. This can be a serious problem that impairs their capacity for water management and flood control. In cases of very high sediment loads in rivers, reservoirs can have relatively short lives due to sedimentation.

Reservoir releases are controlled according to schedules by operators who utilize outlet works to meet water resources management purposes. When emergency conditions occur, as in the case of extreme floods, the spillways are used because outlet works lack capacity to pass the flood flows.

Figure 3.3 shows a simple guide curve for a hypothetical reservoir showing normal high and lower water levels with guidance to the operator to maintain water levels between the two curves. Guide curves can, of course, be more complex and indicate what to do in case of flood and drought conditions.

In hydropower dams, the water elevation above the turbine (H) provides the head for power generation (Figure 3.4). The figure is traced from one posted by

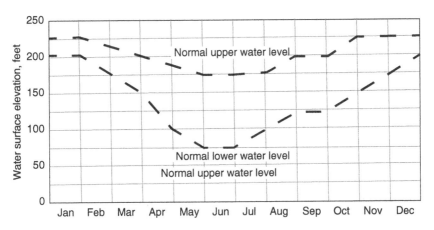

Figure 3.3 A simple reservoir guide curve.

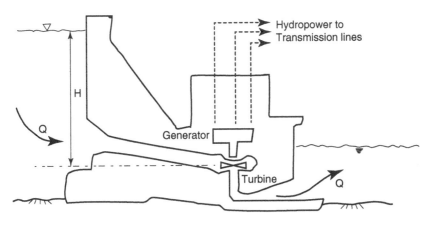

Figure 3.4 Dam cross section with hydropower generation.

the US Department of Energy with credit to the Tennessee Valley Authority, which operates hydropower dams. The kinetic energy or rate of power generation depends on the discharge of water (Q) that passes through the turbine to rotate the blades and generate electric power. Water exits through the draft tube to the tailwater. Energy is transmitted through long distance power lines to electric power networks.

Dams, reservoirs, and hydropower systems pose management issues for water managers and stakeholders. Dam safety is the paramount concern because failure of dams may be catastrophic. Dams should be monitored for geotechnical, structural, and hydraulic condition. Reservoir water quality and eutrophication due to excessive nutrients are concerns in reservoirs subject to pollution from land uses. Water levels can be a source of conflict when lake levels fall below desirable levels for recreation. Environmental degradation occurs from submergence of land and from changed stream regimes. Fisheries can also be affected by changed flow regimes. Loss of salmon runs due to dammed streams is a common example.

Urban Water Systems

Most water infrastructure in terms of value is in urban water systems that provide water supply, wastewater management, stormwater control, and recycled water management. These are separate services, but the concept of an integrated urban water system is emerging to foster coordination among them. The water industry is making efforts to promote this integration through concepts with names like integrated urban water systems and One Water. Daniel Okun, a leading water engineer from the University of North Carolina, did a great deal to advance the concept, which is illustrated by Figure 3.5. Stormwater harvesting can be added to the figure to illustrate a complete picture of the urban water system.

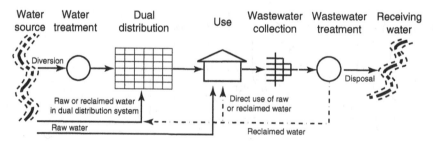

Figure 3.5 Urban water system.

Water Supply Infrastructure Systems

Most water supply systems in urban areas are owned by public utilities, ranging from large business-oriented organizations to small systems that serve only a few people. They provide water services to homes, businesses, factories, and other public and private facilities. Raw and industrial-grade water is delivered to some customers, and sometimes reclaimed water is provided for irrigation and other non-potable purposes. Potable water reuse is practiced in a few situations. The water supply sector also includes many self-supplied end users. Its infrastructure subsystems include source of supply, treatment, and distribution.

The number of community water systems varies with new starts and consolidations, but there are about 50 000 in the United States, and most serve fewer than 10 000 people. The Safe Drinking Water Act (SDWA) defines a community water system as a public drinking water system which serves at least 25 people or 15 service connections for at least 60 days/year. In addition, many non-community water systems, such as campgrounds, highway rest stops, and other transient and non-transient sites, supply water to end users.

Source water facilities range from stream diversions to stored water in reservoirs and to wells and pumps that serve small systems. In some cases, wholesale water may be provided by regional entities and water may be delivered through consecutive systems, by which primary utilities sell treated water to other utilities. Self-supplied domestic users normally require only small quantities of water. A large self-supplier may act as its own utility and be required to seek permission to withdraw water. Bottled water can also be considered as part of the sector. In some countries and in emergency situations, tanker trucks are a basic method of water supply.

To protect water supplies, the SDWA specifies a Source Water Protection program. Examples of available measures include riparian zone restoration to reduce runoff pollution; stream bank stabilization to reduce sedimentation; stormwater controls for agricultural, forestry or urban lands; ordinances for source or wellhead protection areas; emergency response plans; and education about pollution prevention and source water protection.

Treatment systems may be as simple as disinfection at a well or as complex as a large treatment plant with advanced unit processes. A large treatment plant might occupy many acres of land, while a small well and disinfection unit in a rural community may be practically hidden from view. Treatment can also be distributed and handled at point-of-entry and point-of-use systems.

Distribution systems have many pipes, valves, pumps, fire hydrants, meters, and other fittings. In some cases, transmission pipes are used to convey raw or treated water. Transmission and distribution systems in the United States have over two million miles of many types and sizes of pipe. Premise plumbing systems are the "last mile" to provide water to end users in millions of buildings in the United States. The users have responsibility for them, whereas the utility has responsibility for the tap at the main line and the service line portion in the public right-of-way or to a customer meter.

Operation of supply source infrastructure may require sophisticated reservoir controls, compliance with withdrawal permits, and drought management planning, among other measures. Treatment plants require expert operation of unit processes by certified operators and outcomes of treatment must be reported to state regulatory authorities. Operation of distribution systems is becoming more complex due to the scales of the systems, aging pipelines, water quality issues, and scrutiny by the public and regulatory authorities. A certification program for distribution system operators has also been developed.

Concerns of water utility managers vary but normally include aging infrastructure, source water, workforce, regulatory changes, financial issues, and security. The managers worry about access to new supplies and maintaining existing supplies under the threat of climate change. The aging infrastructure problem of water systems is caused by underinvestment in renewal. Water efficiency will be the answer for many supply problems, but conservation may cause revenue shortfalls or rate increases. Concern about security is about sabotage or other threats, whether they stem from humans or natural events.

In higher-income countries, water supply services are expected to work effectively and reliably all the time. In these countries, water supply is changing from a centralized, government-dominated and supply-side industry to a more flexible industry with more demand management and private sector involvement. In lower-income settings, people may simply lack access to safe and reliable water. Here, utilities may struggle to provide service, and lack of adequate financial support plagues attempts at improvement.

Wastewater Systems

Most wastewater service in the United States is provided by government-owned systems and utilities, although operation of treatment facilities is sometimes contracted to private operators. Most utilities are divisions of city governments or special districts. Many types of ownership occur in rural areas, including small businesses, associations, and auxiliary enterprises. Industrial wastewater systems may be attached to centralized systems or operate independently from them. Manufacturing and process industries may be connected to sewer networks and be subject to pre-treatment regulations, or they may have their own discharge permits.

Compared to water supply systems, wastewater systems are difficult to count. They may be embedded in departments of city government and hard to identify as separate utilities. Treatment and collection may be in different management units. USEPA maintains counts of "publicly owned treatment works" (POTWs) and sewer collection systems that indicate about 16 000 POTWs and about 20 000 collection systems in the United States. These totals are of the same order of magnitude as the number of municipalities and special districts providing wastewater service. There are about 750 combined sewer systems in the United States, mostly in the East and Midwest. In these systems, stormwater pipes are connected to sanitary sewer systems.

Wastewater systems serving the public and not connected to networks are like the Transient and Non-Transient Non-Community Water Systems that are regulated in the water supply sector. However, no central statistics are kept on these systems. They might serve factories, schools, campgrounds, stores, rest stops, gas stations and other free-standing facilities and in most cases, they will likely use on-site systems such as septic tanks or package units. If they have large capacities, they may require a discharge permit.

Sources of wastewater are homes, businesses, and industries. Homes discharge domestic wastewater, which tends to be similar and predictable in content. Businesses are as varied as the categories of commercial land uses, and extend across office complexes, food and beverage outlets, hospitals and medical facilities, schools and sports or performance venues, among others. These impose varied demands on wastewater systems, such as those with only customer restrooms to restaurants that discharge waste with large amounts of fats, oils, and greases. Industrial wastewater is a separate category that ranges across types of industries with many unique types of effluents.

The infrastructure required for community wastewater systems comprises collection sewers, lift stations, treatment plants, interceptors and outfall sewers, and sludge processing and disposal facilities. Collection systems begin with local service lines, then collector sewers, and then trunk sewers. They may receive stormwater, either by design as in combined sewers or by accidental connections. They may also be subject to infiltration when buried pipes lack integrity. For maintenance purposes, stormwater connections and infiltration are dubbed I&I, for inflow and infiltration.

Wastewater collection systems normally operate by gravity. Due to the many different constituents of wastewater, blockages occur even in the best managed systems. How the systems are maintained is often the main issue in operations. USEPA has developed a Capacity, Management, Operation, and Maintenance program (CMOM) to organize the activities required, like periodic inspections to detect the buildup of blockages.

Wastewater treatment operations involve unit processes with physical, chemical, and biological systems. Regular reports to regulatory authorities of their performance are required. The treatment systems range from simple lagoons to complex combinations of unit processes. Wastewater can sometimes be recycled, which usually requires additional treatment. Sludge that results from wastewater treatment must be managed separately from the liquid effluents discharged from treatment plants. Rural areas and industries not served in urban areas usually use on-site systems and package plants.

Managing wastewater systems is challenging due to their invisibility or being "out of sight, out of mind." This may cause neglect, under-investment, and lack of maintenance.

Obsolescence and wear and tear may cause old systems to fail, causing difficult community and environmental problems. Risks involved include the failures, regulatory violations, pollution, disease outbreaks, and sabotage. Failures have large impacts on the environment. Finance can be a difficult problem. Investment is needed for renewal and modernization and operating budgets must be adequate. However, ratepayers are often reluctant to provide adequate funding.

Adequate sanitation provides essential barriers that protect vulnerable people from waterborne diseases such as cholera, typhoid, and dysentery. However, wastewater management is especially difficult in low-income countries where access to sanitation and wastewater service is often lacking. Few wastewater treatment plants operate in many of these countries.

Recycling of wastewater offers opportunities to advance water efficiency and address problems of scarcity. Recognizing this opportunity, wastewater utilities may consider expansion of their missions to become "utilities of the future" that handle total recycling of materials, such as food wastes.

Stormwater

The main function of stormwater systems is for drainage of sites in cities and developed areas, such as roadways, industrial plants, and others. Stormwater quality requirements were added to the drainage function under the Clean Water Act. Systems range from simple site drainage systems to those in complex megacities.

The configuration of stormwater systems begins with infrastructure components to capture runoff, such as gutters and inlets (Figure 3.6). Systems are classified as minor or major depending on the rates of flow to be managed. Minor systems will handle frequently-occurring drainage, whereas major systems are for larger and rarer flood flows. Minor systems are often equipped with treatment devices, such as filters and settling ponds. Currently, green infrastructure systems such as rain gardens and permeable pavements are being installed on a widespread basis and stormwater harvesting, treatment, and recycling are practiced increasingly. Stormwater is ultimately transmitted to outfalls in receiving waters.

Stormwater systems function mostly by gravity, but pumping can be used where needed, such as to drain a detention pond more rapidly than gravity alone.

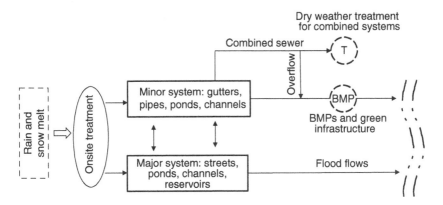

Figure 3.6 Stormwater system.

Maintenance is important because drainage systems collect leaves and other debris, which can clog inlets and render systems inoperable.

Stormwater mobilizes many pollutants, and treatment of non-point runoff has become important with urbanization, which exacerbates both quantity and quality problems. If stormwater systems are not designed and operated effectively, they can degrade ecosystems. For example, a culvert can block movement of natural species along a small stream. The land use changes that accompany stormwater development may destroy natural wetlands. The combination of such effects is responsible for much loss of biodiversity.

In large cities, governance of stormwater systems can require intergovernmental coordination as systems may be controlled by different local governments who must work together to ensure unobstructed and effective drainage to major channels. Lack of stormwater services may create environmental justice issues that include ponding, mosquito breeding, and trash buildup. Financing stormwater can also be difficult and many communities have developed stormwater utilities, but there can be opposition by groups that claim that stormwater fees are taxes in disguise.

Irrigation and Drainage

Altogether, the irrigation sector accounts for about 70% of total water global withdrawals and for more consumptive use than any other water use. As world food demands increase, it will continue as the major water-using sector, but with changes that increase the efficiency of water use.

Irrigation water is essential for crop production in dry regions, and in humid areas it can be important for supplemental water to improve crop yields, especially during drought periods. The scale of application ranges from small-holder subsistence farms to large farm enterprises. Water is also used on farms for livestock and aquaculture. Irrigation is connected to the social fabric of societies in the same way as banking, schools, and other core services.

In the United States, irrigation practices are well established, and changes are likely to be incremental. With urbanization, new subdivisions move into farm regions and purchase water supplies from irrigators. Urban farming is also on the rise and requires support from water utilities as well as to develop wells and some settled areas. Expanding urban areas also require irrigation for landscaping.

Infrastructure systems to support irrigation and drainage range from large dams and canals to microlevel systems. The main types of irrigation systems are surface (gravity, flood, and piped), sprinkler, drip/trickle, and subsurface. In a comprehensive system, a storage reservoir can feed main canals that distribute irrigation water to laterals. A main canal can feeds an uphill pressure pipe system using a pump or a gravity system on the downhill side. A large canal can include a diversion dam to feed the headworks of a smaller canal that might serve a center pivot system and/or a flood irrigation system. Additional methods like sprinkler systems, automated pressure pipe systems, and drip and subsurface systems can also be involved.

Compared to M&I water, the organization of irrigation operations is more diverse. While some are within organized management systems, but many operate as part of user associations, mutual companies, and other informal arrangements. Management systems to allocate water fairly among upstream and downstream users are required. Effective management arrangements are often lacking and depend on cooperation among users. Techniques such as monitoring, automatic control, and systematic maintenance are sometimes difficult without effective management systems.

In the future, irrigation will face rising competition for water supplies and there will be opposition to new projects. Sometimes there will be reallocation of irrigation water. The trend is toward smaller, more efficient systems and smart micro-irrigation and intelligent controls. Governance should empower farmers at different levels and facilitate access to irrigation water, including increased uses for urban farming. At the same time, it must respect the need for farmers to be independent and respond to market conditions.

Finance to sustain and renew irrigation systems is difficult and enterprises are not always able to support infrastructure systems through revenues of farming. Varied forms of funding are used such as user charges, government subsidies, and cooperation with multi-purpose projects. Environmental and health issues will be important in the future and will require feasible approaches for regulation of the quality of irrigation return flows and to protect ecosystems.

River and Coastal Works

River and coastal works include infrastructures such as levees, hydraulic training structures, and erosion controls. Flood control reservoirs and levee systems should work together for integrated flood management. Coastal works, such as sea walls, groins, and breakwaters, can be used as sea level rise continues. River training structures are controversial due to environmental effects, but many navigation systems depend on them to maintain channels. Flood control reservoirs may be complemented by locks, re-regulation dams, and various river training structures such as dikes and weirs. Levees are often constructed along the banks of streams to prevent flood waters from reaching structures and farmlands.

From an environmental standpoint, river and stream systems should ideally be left in their natural conditions. This is only possible in a few situations, and most streams have been modified with artificial controls, including dams to regulate the flow. Modifications in flow regimes such as intentional surges to mimic natural flooding may improve environmental conditions.

Coastal works are essential in many places to mitigate damage from storm-induced surges, and the problems will increase with sea level rise due to climate change. They include seawalls to be built along beachfronts for direct protection against tidal and wave action. Groins are perpendicular to the shore and used for erosion control. Breakwaters are used offshore and parallel to beaches to dampen wave actions.

Managing river and coastal works is challenging due to the dynamic nature of weather and hydraulic conditions. Also, environmental needs often conflict with economic solutions for flow control. Sedimentation and erosion control are major concerns in many places. Improved runoff forecasts and reservoir operational procedures are needed to minimize damage and balance objectives. Vulnerability of levees is of increasing concern due to population growth and uncertain maintenance performance.

Canals and Pipelines for Water Conveyance

Water is heavy and more difficult to move long distances than electric power, but transporting it is often necessary because water uses do not always occur near the sources. Transport requires transmission and distribution infrastructure involving both closed conduits and open channels. Canals and large pipelines are often used to move water over long distances, whereas large and small pipelines are involved in urban water systems with shorter distances in most cases.

The pipelines in urban areas form a hierarchy from transmission lines that transport water from source to plant or from plants to distribution systems, to distribution mains and lateral pipelines that distribute water around a community, and then to service lines which are small diameter pipes from distribution mains to user plumbing systems. Also, in – plant piping located in pump stations or treatment plants can be extensive. Wastewater systems have a parallel classification, including collection systems, laterals, trunk sewers, and outfalls. Stormwater systems involve similar piping.

Maintenance and renewal of pipeline infrastructures are a concern because they are often "out-of-sight and out-of-mind" and can be neglected. This creates buildups of deferred investments, which can add up to substantial needs for renewal. This is particularly true with urban distribution systems, which exhibit massive levels of deferred renewal needs in most cities.

Natural Water Infrastructure

Water infrastructure plans should include consideration of the contributions of natural systems, including aquifers, wetlands, floodplains, and natural streams. Managing these requires a mix of measures beginning with recognizing the value of their ecological services and carrying through to development of definite management plans.

In particular, use of groundwater from aquifers offers substantial opportunity to avoid building new infrastructure and can help to manage natural conditions. The infrastructure for wells includes their basic construction along with pumping and distribution facilities. Aquifer storage and recovery systems are increasing in use as we learn to manage aquifers better.

Small Water Infrastructures

While large infrastructures are in the spotlight more than small systems, the "last mile" of systems like premise plumbing are significant in water services. Premise plumbing for drinking water begins at the service line and continues into buildings. It involves pipelines, valves, meters, and sometimes pumping systems. The quantity of these small systems is very large and they have many issues, like for example legacy lead service lines, which pose continuing health risks. By the same token, small systems are found for wastewater, stormwater, and irrigation systems, and there are many small dams as well as hydraulic works in small streams.

Broad Issues of Water Infrastructure

Water infrastructure is a diverse category and general issues are difficult to characterize, but management, finance, and risk reduction are common themes. Managing water infrastructures can be improved through incorporation of formal asset management systems. These are needed to respond to the ongoing deterioration of systems, especially those that are not visible but critical to sustain society, like buried pipelines.

Investment needs to sustain water infrastructure are enormous. In high-income countries these needs address the requirement to renew legacy systems that have deteriorated. In lower-income countries, massive investments are needed for new infrastructures. Financing mechanisms depend on whether the systems are public or private. Many dams, hydro plants and river works are built by governments, but dams and hydroelectric projects are also built by privately owned utilities and investors. Drinking water and wastewater systems are normally built by local governments and should be funded by rate payers. Many private water systems also operate.

Irrigation systems are built and managed by a range of operators, including governments and cooperatives, as well as private operators. User charges normally do not pay the whole costs of systems.

Public–private partnerships for water infrastructure systems are increasing, to include outsourcing of public utility support services, contract operation publicly owned treatment works, design-build operate contracts, sale of government-owned assets to private companies.

Failure of non-resilient water systems is among the highest global risks. Most water systems around the world are not resilient and do not meet basic needs. Given the risks of water system failure, security has risen on the scale of importance globally. Security of water supplies for people and agriculture is needed as a buffer against scarcity and the vagaries of nature. Dam safety and flood security are required as safeguards against loss of life or property where rising populations in harm's way are increasing risks to events such as levee failure. Cyber security is necessary because water infrastructure systems are vulnerable to hacking and the consequences of malevolent behavior can be severe.

Questions

1 Explain the problem of cyber security for water infrastructure. Give an example.

2 What is the main difference in the status of water infrastructure systems of high-income countries and low-income countries?

3 Give an example of how a dam's characteristics would be described in terms of construction method, height, purpose, and ownership.

4 If dams have major impacts on natural systems, why are many of them in use?

5   List the main storage zones of reservoirs for purpose of water management operations.

6   What are differences between a storage reservoir, run-of-river reservoir, and dry dam?

7   What are possible failure modes of dams?

8   What are the functions of the two spillways of many dams?

9   What are the main difficulties encountered when removing old dams?

10   What is the difference between a run-of-river reservoir and a storage reservoir?

11   Make a sketch and explain how pumped storage works.

12   What is difference between KW and KWH?

13   What is largest hydro plant in the world and approximate generation capacity?

14   What is the main benefit of a hydro dam when it is combined with other forms of electric generation?

15   Does a wastewater collection system usually work by gravity or pressure?

16   Explain some ecological impacts of culverts.

17   Explain the different kinds of benefits provided by stormwater systems.

18   Explain who owns service lines and has responsibility to maintain them.

19   How does a combined sewer work?

20   How does a detention pond work?

21   Is a "community water system" the same as a water supply utility? Explain.

22   Is recycled wastewater potable?

23   Name two unit processes of a water treatment plant.

24 On the basis of risk, explain the difference between a minor and major stormwater system.

25 Sketch a plan view of a typical wastewater treatment plant and identify the main functions of its components.

26 Sketch a plan view of a typical water treatment plant and identify the main functions of its components.

27 Water Supply Systems are normally owned by local governments, utilities and private companies. How does this ownership differ for wastewater systems?

28 What are SSOs and CSOs in wastewater?

29 What are the main functions of a storage tank in a water distribution system?

30 What are the purposes of force mains, pressure sewers, and lift stations for wastewater management?

31 What are the three main pipe materials used in urban water supply systems?

32 What are three important unit processes of a wastewater treatment plant?

33 What is the function of a constructed wetland and how does it achieve its purpose?

34 What is the nature of the problem of aging distribution systems?

35 When is desalination a good choice for water supply?

36 Where is desalination practiced the most and what is the water used for?

37 Why is governance of stormwater systems challenging in large metropolitan areas with different jurisdictions?

38 Why is wastewater infrastructure complex and expensive?

39 Why is there no centralized inventory of all irrigation systems as there is for water supply?

40 Name a micro-irrigation technology and explain how it works.

41 Name three different types of irrigation systems.

42 How does a lock work? For upstream navigation? For downstream navigation?

43 What are some ecological effects of locks?

44 Sketch a plan view of a river discharging into the ocean and identify the following types of structures: locks, levees, training walls, diversion dams, erosion controls, and jetties.

Problems

Power and Energy Equations

The basic equation of water power generated by a turbine is:

$$P = \eta \gamma Q H,$$

where η = efficiency, γ = specific weight, Q = discharge, and H = head.

In the English system, the usual equation for power of a turbine gives the result in horsepower (HP):

$$HP = \frac{\eta \gamma Q H}{550}$$ For a pump the efficiency term is in the denominator,

$$HP = \frac{1}{\eta} \frac{\gamma Q H}{550}$$

$1\,HP = 0.746\,KW.$

In the English system the result would be feet-lb/second and in the SI system it would be watts. In the SI system, use $9.81\,kN/m^3$, Q in cms, and H in m.

Power of a Turbine

What is the power generated by a turbine with a flow rate of 10 cfs, a head of 50 feet, and an efficiency of 80%? (HP = 0.8 * 62.4 * 10 * 50/550 = 45.4 HP (or 33.9 KW).

Horsepower of a Pump

What horsepower pump is needed to pump a flow rate of 5 cfs against a head of 100 feet with efficiency = 80%? (HP = (1/0.8) * 62.4 * 5 * 100/550 = 70.9 HP).

Energy Generation

What is the energy required to operate the 70.9 HP pump for 24 hours? (Energy = 70.9 * 0.746 * 24 = 1270 KWH).

Hydroelectric Energy

A large reservoir has an average water depth at the dam of 250 feet. Assuming a generation efficiency of 80%, estimate the average annual discharge in cfs required to produce 7500 million KW/h of hydroelectric energy in a given year.

Examples of Units

A bottle contains 750 ml of water. How many fluid ounces is that? (1 gal = 128 oz = 3.7854 l = 3785 ml. 750 ml = (750/3785) * 128 or 25.36 oz).

If a treated water storage tank in a city zone holds 2 MG, how many acre-feet is that? (Volume = 2 million gallons * 3.0689 = 6.14 acre-feet or 2.001 MCM).

If when half-full, the contents of a large reservoir are 200 000 acre-feet, what would this be in MCM? (Capacity = 200 000 AF = 247 MCM).

A reservoir has a capacity of 10 000 acre-feet. How many gallons is that? How many meters^3? (10 000 AF = 3.259 billion gallons = 12.33 MCM. Note that AF and MCM are used for large quantities of water, with 1.0 MCM = 811 AF).

A household tap runs at 3 gallons per minute (gpm). How many cfs is that? How many l/s? (3 gpm/448.836 gpm/cfs = 0.00668 cfs. 5 gal * 3.7854 l/gal = 11.356 l/minute or 0.189 l/s).

A pump in a municipal water supply system is rated at 200 gpm. How many cfs is that? If it pumps for 24 hours, how many acre-feet has it pumped? (Pumping = 200 gpm/448.836 gpm/cfs = 0.446 cfs. For 24 hours it pumps 38 500 feet^3 or 0.884 AF).

The annual average pumping rate for a well is 500 gpm. Convert this to cfs and find the volume pumped in a year in acre-feet. (Rate = 500 gpm/448.836 gpm/cfs = 1.11 cfs Vol = 1.11 cfs * 365 days = 406.6 cfs-days = 807 AF).

A stream flows with 10 cfs for five days. How many cfs-days is that and how many acre-feet? (10 * 5 = 50 cfs-days = 99.18 AF. Note use of the unit cfs-days, which measures volume. 1 cfs-day = 1.9835 AF.)

Units

A river flows with an annual average of 50 000 cfs. How many cubic miles will it discharge in 10 years? 50 000 * 723.97/(3.3792 * 1 000 000) = 10.7 miles^3/year or 107 miles^3 in 10 years).

If the Amazon River has an annual discharge of 7.2 million cfs, what will be the flow in km^3/year and MAF/year? (204 038 km^3/year or 5212 MAF/year).

4

Demands for Water and Water Infrastructure

Introduction

The demands for water and use of water infrastructure determine the required dimensions, capacities, and operational responsibilities for facilities such as reservoirs, treatment plants, and pipelines. Sometimes demand refers to a quantity of water that might be influenced by price, sometimes for use of infrastructure like a reservoir, and sometimes as a fixed requirement like a pollution control mandate. In this chapter, the word demand is used to cover these cases.

This chapter explains how these demands depend on uses of water and the need for services like wastewater management or flood risk reduction. It provides information to make order-of-magnitude estimates of demands. Demands are introduced and explained for each water management sector, but many details are beyond the scope of the chapter. The reader can find these details in the literature of the sectors, like demands for urban water supply, regulations for pollution control, and standards for flood control, for example. There is abundant discussion of the demand for residential water supply, for example, but the literature on commercial and industrial water uses is more dispersed and difficult to summarize. Irrigation engineers have defined the demands for water in agriculture with extensive explanations, equations, computations, software, and tables that are available in professional references.

Demands and Requirements

Demand studies should answer questions such as "how much water supply should the utility have," or "what should be the capacity of this treatment plant, pump station, or other element of water infrastructure?" The analysis will

Water Resources Management: Principles, Methods, and Tools, First Edition. Neil Grigg.
© 2023 John Wiley & Sons, Inc. Published 2023 by John Wiley & Sons, Inc.

normally consider the differing requirements and demands for urban water supply, industrial supply, wastewater, environmental water, stormwater, irrigation, flood control and instream flows for hydropower, navigation, and recreation.

Demand can measure how much is required or how much is simply requested. Examples of requirements are a crop needing 20 inches (510 mm) of irrigation water in a season to produce a certain yield and a person needing at least 10 gal (38 l) of water/day for a minimum level of hygiene. An amount requested could be for a discretionary use, like a homeowner using a large quantity of water to keep a lawn beautiful. Requirements for some purposes, such as pollution control, are set by legal mandates and do not offer choices to management organizations. On the other hand, choices for urban, industrial, and irrigation water supply exhibit mixtures of requirements and demands. An example is residential water supply where a family will require a certain base amount of water, but they may demand additional water for purposes such as lawn irrigation or swimming pools. The water manager will need to know whether a demand is a requirement set by law or is discretionary with a variable component. Sometimes even minimum requirements cannot be met, and plans are needed to balance competing demands when all of them cannot be met.

Direct requirements and demands for water supply and water management actions are determined by social and economic factors and are served either by intermediate agents, such as water authorities, or by self-supply systems. In turn, these determine the requirements for water infrastructure capacity. The agents may own and operate facilities that meet multiple demands through shared use of infrastructure and multipurpose water resources management. For example, a reservoir may store water for cities, farms, and hydroelectricity producers, while its storage space may also provide flood control services. When water facilities are used this way for more than one purpose, demands may converge and cost shares will be allocated to different parties, like to cities for water supply and to flood control authorities for use of the same infrastructures.

Water Demands by Sector

Water demands can be organized by sectors that reflect the purposes of the water uses or services provided. Viewing them by sectors enables us to see how trade-offs are needed to balance water uses across the sectors. It is easy to get confused about sectors because the word is used in different ways. It means the divisions of something and is closely related to the word system. General sectors are those like society, the economy, and the natural world. These divide into more specific sectors like the natural world dividing into land, water, air, and living things. Food and energy are sectors that support society. Sectors exhibit demands for

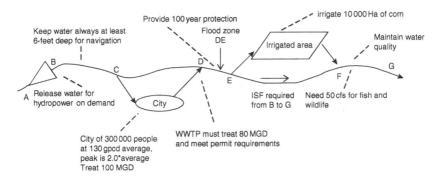

Figure 4.1 Demands for water management along a regulated stream.

water in different ways. For example, water management for purposes like drinking or irrigation can serve multiple sectors, such as the urban or industrial sectors.

Demands normally begin with water supply for cities and industry, which involves infrastructure for water diversion and storage, treatment, and distribution. Wastewater demands create need for capacity in collection sewers and wastewater treatment systems. The agricultural sector needs water for crop production, aquaculture, and livestock. In-stream demands include environmental water, hydropower, and navigation. Flood control and stormwater demands focus on use of facilities such as conveyance channels and water storage.

In watersheds, water is used to meet multiple demands from different sectors along a stream reach. As Figure 4.1 shows, water releases for hydropower determine reservoir operation at the upper end of the reach. The figure shows representative values for different demands along the stream. The releases are constrained by a depth requirement for navigation and by demands for water by the city as well as for flood control. Demand for stream water quality will drive capacity of the city's wastewater treatment plant, and enough water for fish and wildlife may be required by instream flow rules, along with those for environmental water quality. The irrigated area may also have rights to water to meet its demands. The demand for flood protection at vulnerable points along the stream will be expressed in terms of risk factors.

Water Use Data

When studying a particular situation, local water use data is required to estimate future water demands. At the national level, water data are compiled by US Geological Survey (USGS), which has developed a classification system and has published data every five years since 1950. These data are available from the USGS

Table 4.1 Estimated water use in the United States (MGD).

	1995	2000	2005	2010	2015
Population, millions	267	285	301	313	325
Public supply	40 200	43 300	44 200	42 000	39 000
Domestic use	3 390	3 720	3 830	3 600	3 260
Commercial	2 890	NR[a]	NR	NR	NR
Irrigation	134 000	137 000	128 000	115 000	118 000
Livestock	5 490	1 760	2 140	2 000	2 000
Aquaculture	NR	3 700	8 780	9 430	7 550
Industrial[b]	22 400	19 780	18 190	15 986	14 786
Mining[b]	3 790	3 500	4 020	5 320	4 000
Thermoelectric[b]	189 900	195 500	201 100	160 900	132 900
Total use	402 020	408 000	410 000	355 000	322 000

[a] NR = not reported.
[b] Industrial, mining, and thermoelectric include some saline water. In 2015, these were 5, 53, and 28% respectively.

website. While they cannot be used directly in local situations, they indicate trends and relative demand levels that can be useful.

Table 4.1 shows historical data for the five most recent USGS surveys and can be analyzed to show trends. The categories provide detail on urban and industrial supply and irrigation water use. The data show that, despite rising population, total water use has been declining, particularly public supply, irrigation, industrial water, and thermoelectric cooling water.

United States Water Balance

A national water balance can place these water uses in perspective. Interpreting the data requires specification of withdrawals and consumptive uses because non-consumptive uses return water to streams and a micro-view is needed to see the exchanges between one form of water and another. A water balance diagram (Figure 4.2) illustrates this for all US water use data. The diagram is an input–output display with detailed breakdowns. It shows sources of surface and groundwater and uses in four groups (domestic-commercial, industrial-mining, thermoelectric, and irrigation-livestock). Public supply is an intermediate category and transfer mechanism from source to use. The consumptive uses of water and return flows

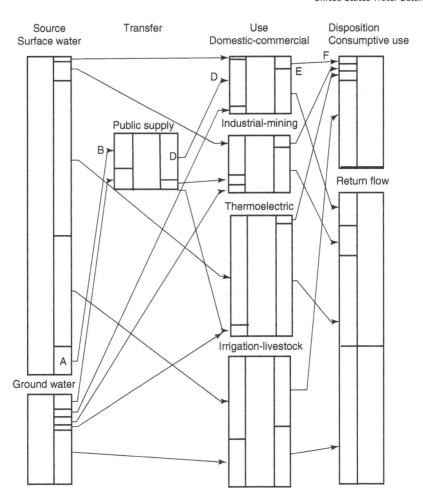

Figure 4.2 Distribution of water withdrawals and consumption in the United States.

are also shown. Most thermoelectric water is returned but most irrigation water is consumed. This complex diagram was included in the 1995 USGS Water Use report, and it has not been included since then. How to interpret the data for the diagram is explained below.

The data has been deleted from the figure to remove out-of-date numbers and to present the methodology with more clarity. The labels are the same as on the original diagram except that the term "transfer" was added above the public supply rectangle. Each rectangle contains a percentage of water and in some cases the total amount in MGD. These numbers would change with each report. To understand the data, follow the arrows shown as follows:

A–B: Quantity and percentage of water from surface supplies to public supply and quantity and percentage of public supply coming from surface supplies

C–D: Quantity and percentage of water from public supply to domestic-commercial use and quantity and percentage of domestic and commercial use coming from public supply

E–F: Quantity and percentage of domestic and commercial use being used consumptively and percentage of consumptive use by domestic and commercial use

In Chapter 6, a diagram of the full US water balance is shown. By combining the information from it and Figure 4.2, you can get a comprehensive view of the national hydrologic balance.

Categories of Water Use

Of the USGS water use categories, the water manager needs estimates of public supply, irrigation, and self-supplied industrial water most frequently. Some industrial supply is included in public supply, so it can be combined with self-supplied industrial water to estimate the total industrial water use in a community.

If information about the other categories are needed, the USGS data can be combined with data about them to make the estimates. For example, self-supplied domestic use for individual homes can be studied from Census data about urbanization and data about numbers of small community water systems. Livestock and aquaculture water uses can be estimated from the number of animals and economic activity. Mining can be lumped in with self-supplied industrial uses, and thermoelectric power use can be analyzed from energy data.

Data on public supply can help in making estimates of urban water demand. On Figure 4.2, public supply is an intermediate category for distribution of water to end uses like domestic and commercial uses in cities. As an example, the diagram shows how part of public supply goes to domestic and commercial uses and how much was consumed. It also shows water uses from public supply to industrial and mining and to thermoelectric cooling. Part of the total public supply is for public uses, such as pools, parks, firefighting, water and wastewater treatment, and municipal buildings, and some is unaccounted for due to leaks, losses, and maintenance.

In its 2015 study, USGS estimated that 60% of the public supply withdrawals served domestic uses. This showed a use of 82 gallons per capita per day (gpcd) (310 l), based on 60% of 39.0 billion gallons/day for a population served of 283 million. The remainder of the 2015 population of 325 million was self-supplied and showed a use of 77 gpcd (291 l). These average values vary by the situation and should not be used for planning purposes. Plans for individual communities require studies of historical use patterns, demographics, and trends in commercial and industrial demands.

Most domestic and commercial water supplies are distributed by utilities or Community Water Systems. The United States has about 50 000 of these and most serve small populations (see Chapter 12). In some cases, a utility may sell wholesale water to another utility and the quantity will be included in the withdrawal. Managers in utilities estimate future water demand to plan expansion of source, treatment, and distribution facilities. The estimates normally will be for daily average, daily peak, and hourly peak. The system requirements for water supply facilities must consider time-varying needs, reliability of infrastructure, the response of users to regulations and incentives, and risk of shortage.

Source of supply (stream, reservoir, or aquifer) should include reserves and be capable during drought. Some utilities might have a rule of thumb such as the source must deliver 150% of average demand, but a better approach will consider the historical statistical record, which will reflect climate, land uses, types of buildings and socio-economic status. Chapter 5 explains how low flow statistics can be used for such estimates.

Treatment facilities must meet varying demands and have redundancy. Demands in each locality will depend on conditions such as the employment base and population demographics. In a bedroom community, demands might show two peaks to correspond to the working day, while a community with more commercial uses and retired population might see demand spread uniformly during the day.

Water distribution systems must meet all demands including fire protection during outages, such as main failures. They usually have limited in-system storage, so the network design must consider contingencies such as location of critical facilities, redundancy, and emergency operations.

Customer classes for rate studies (see Chapter 19) can be used to categorize demands. In cities, customer classes include residential, commercial, industrial, wholesale, and public. Each use within a customer class can be broken into components. For example, residential uses include drinking, toilet and bathing, kitchen, laundry, car washing, and outdoor uses.

Data from billing and from metering can be used to quantify demand. Chapter 10 explains the process of auditing water uses, where authorized uses and losses sum up to total water production. The authorized uses can be metered or unmetered and billed or unbilled. Metered and billed water will be the major component of revenue water, which provides the funding stream for the water utility. Revenue water comes from residential, commercial, government, industrial, and wholesale customers. Public uses such as fire-fighting, irrigation of public spaces, and flushing of water mains are normally unmetered and unbilled.

While per capita domestic use in the United States in 2015 was about 82 gpcd (310 l), if everyone lived in apartments, without any outdoor use of water, the per capita demand would be lower due to less outdoor use. Commercial and industrial uses in cities are diverse and difficult to classify. Commercial water use can be related to types of businesses and by sectors. Industrial uses from distribution system water will depend on specific situations and are difficult to generalize. They continue to change, and many high-water-use industries have moved to countries where they can access resources more easily. Of course, wholesale deliveries of water are unique to individual situation. Public uses also vary among cities, but historical data should be useful to estimate them.

Predictions of urban water demand can be simple, such as a graph of past water uses projected into the future, or they can use complex statistical and socio-economic models. The research literature to support models has grown substantially, but most practical applications are based on studies in individual localities.

A model developed during the 1970s named IWR-MAIN (the acronym for Institute of Water Resources, Municipal, and Industrial Needs) provided a framework to display the main variables, but it is no longer available and its owner now provides customized spreadsheet-based models for its own specific client situations. While the model is not available, the underlying reports and research provide insight into the drivers of water demand.

IWR-MAIN disaggregated urban water use determinants and the uses could be estimated from multipliers based on data in urban areas. The residential sector used Census categories such as single- and multi-family housing data and the estimates considered drivers of water use such as income level, household size, and other demographic variables. Non-residential use was simulated by industry codes and considered employment and productivity. Water managers can develop forecasting models using similar concepts for residential, commercial, and industrial water uses in individual settings.

For a simple example with hypothetical numbers, a water manager might require an estimate of future urban water demand for a city with present population of 125 000. The gross per capita use might be, say, 150 gpcd (568 l) for a total annual use of 6.84 BG (25.9 MCM). A study of local metering data shows that residential uses were 100 gpcd, or a total annual use of 4.56 BG. The difference of 2.28 BG is industrial, commercial, and public uses. The estimate is that population will grow to 150 000 in 20 years, but per capita residential uses will drop to 90 gpcd due to conservation pricing. This means that total residential uses will be 4.93 BG. If other uses remain constant, total annual urban demand will now be 7.21 BG and gross use will be 132 gpcd, down from 150 gpcd. It is also possible that peak demand will change due to the shifting demographic profile of the city.

Irrigation

In irrigated regions, watershed and river basin studies require estimates of agricultural water uses. The problem might focus on water requirements or how irrigation uses impact stream systems and other water users. Irrigation is also used in some humid regions when little rainfall is available seasonally, and it is also used for developments such as golf courses, landscapes, and aesthetic purposes. Agricultural demand may also include livestock and aquaculture, but these will be minor compared to crop irrigation.

In the western United States, the rising demand for water has stimulated interest in the transfer of water rights from farms to cities. Large irrigators can also partner with cities in city-farm water sharing such that farmers can idle land in dry years so cities can have water. Cities may also experience increased demands for water systems for urban farming.

Irrigation uses can be very diverse. Large irrigators include farms with surface irrigation systems, wells and center pivot systems, and golf courses. At smaller scales, irrigation may be by wells or from collective systems organized by groups of farmers. Smaller irrigators might include greenhouse operators or small farms with wells or pumps in streams. Irrigation in urban areas to provide water for local food production or neighborhood vegetable plots may come from wells or from distribution systems. The irrigation sector also includes aquaculture with fish ponds and water-handling facilities. Water is also needed for livestock as farmers may use the same water-handling facilities as for crops. A cattle operation might irrigate feed crops and use the same water for cattle in the range or in feedlots.

As water becomes scarcer, economics drives movement of irrigation water from lower-value crops such as hay and corn to higher-value crops such as vegetables and turf grass. Also, more efficient use of water through sprinkler and micro-irrigation systems is evident. Management of crop water use is becoming more scientific through smart water management. This requires more focus on agricultural water accounting to explain losses through various phases of the water transport and use systems.

Quantitative demand for irrigation water is determined by crop requirements. Calculating demand is based on equations for evapotranspiration, which have been studied extensively. References provided by organizations such as the California Department of Water Resources and the Food and Agriculture Organization of the UN are useful. The California Irrigation Management Information System (CIMIS) operated by the California DWR has automated weather stations to provide information to irrigators. It uses the Penman–Monteith equation to estimate evapotranspiration based on weather, soil, and plant parameters. The approach is based on a reference ET based on well-studied grass or alfalfa surfaces and standardized surface conditions. A crop factor or crop coefficient (K_c) is used to calculate the ET_c for a specific crop in the same zone as the weather station.

Crop evapotranspiration, ET_c, is calculated by multiplying the reference evapotranspiration, ET_o, by the crop coefficient, K_c:

$$ET_c = K_c\, ET_o,$$

where
ET_c crop evapotranspiration (mm/day),

K_c crop coefficient [dimensionless],
ET_0 reference crop evapotranspiration (mm/day).

Industrial and Thermoelectric Water Uses

Industrial water use can be very small or large and depends on the situation. In the United States, self-supplied industrial water is about half of the total withdrawn for all public systems. Common industrial water uses are for sectors such as chemicals, food and beverage, pulp and paper products, steel, electronics and computers, metal finishing, petroleum refining, and transportation equipment. In chip manufacturing, cleaning and rinsing can require a great deal of water. Apparel can also be water-intensive, such as cotton production and textile processing.

No general guidelines will apply, and each case must be evaluated individually. Published guides will show large fluctuations, such as in the 2009 World Water Assessment Programme report that shows paper production use varying from 21 000 to 528 000 gal (2 ml)/ton of product.

Thermoelectric power water withdrawals depend on the situation as well. The electric power industry uses water for cooling and emissions scrubbing in fossil fuel plants and nuclear power plants.

In developing countries adequate public water supply infrastructure and regulatory controls may be lacking and self-supplied industrial water uses by unregulated industries may cause problems such as groundwater declines, land subsidence, flooding, and pollution.

Wastewater Management

Wastewater management includes the linked services of sanitation, sewerage, and wastewater treatment. Environmental water quality management is explained subsequently wastewater management. Combined sewers are a special case that combines wastewater and stormwater.

The wastewater sector has moved toward the utility model with service provided by local utilities and paid for by customers. Demand is generally mandated and fixed, rather than based on customer choice, because the quality of wastewater is regulated. In the United States, the Clean Water Act provides the basic set of rules. Some countries may have rules but no enforcement. In many cases separate authorities handle collection networks and treatment plants.

Sanitation means access to facilities for human waste disposal and hygiene. Wastewater service begins with sanitation at user facilities where it is the responsibility of the homeowner or business to provide local access to facilities. No national regulations address demand for toilet facilities, but plumbing codes specify local requirements.

Often sanitation facilities are not connected to sewers, which is the case in many rural areas and developing countries. Some on-site systems, like septic tanks, can provide effective service, but they may malfunction and degrade environmental water quality through soil contamination and nonpoint source runoff.

The collection of wastewater requires systems with many miles of collector and main sewers. For residential customers, the requirement is to collect and remove the wastewater. Commercial and industrial customers generate different strengths of wastewater and may use pretreatment systems to reduce demands on public wastewater systems. Combined sewer systems are a problem in some cities.

Demand for service is based on required capacity and depends on the volume of wastewater generated. A typical minimum diameter for a wastewater line (lateral) to a building is 4 inches. In some locations and in older installations, smaller pipes may be in place. For multi-family, commercial and industrial buildings, the size will depend on the wastewater volume, which may be estimated from the fixture units. Fixture units are a tool to estimate requirements in plumbing design for water supply and wastewater systems in buildings.

People can influence the quantity of wastewater they generate through water conservation or gray water systems, and industries can choose or be required to pretreat their wastewater to reduce their loads on the collection system.

Treatment plants vary from small onsite systems to advanced WWTPs, which are complex to operate. Demand for treatment is determined by the regulatory controls, which are correlated with stream standards under the Clean Water Act. Requirements will be specified in the discharge permits.

Stormwater Management

Demands and requirements for stormwater systems address minor drainage, urban flooding, stormwater quality management, and any additional needs established by local governments such as, for example, use of stormwater systems to enhance open space or ecological conditions.

Minor drainage requirements are established by criteria, such as a two- or five-year storm, for example. This will depend on local requirements.

Demand for land drainage depends on local rules and politically determined levels-of-service. Land owners want their property drained, no matter how much rain occurs but that demand must be tempered by the cost involved and the trade-offs in supply of the service, as well as whether "pretreatment" methods such as low-impact development are implemented.

Demand for stormwater quality control is determined by regulatory rules rather than customer choices. Demand for control of CSOs is a hybrid among wastewater and stormwater. Urban stormwater services may also include urban flood control, which is also required in rural zones.

Flood Control

Demand for flood control is like stormwater drainage where a person wants security against flood risk, no matter how extreme the event. Flood control water management can be structural (dams and levees) or non-structural (zoning, insurance, warnings, etc.). Demands for these measures are variable and difficult to characterize. Normally, a requirement will be set by some policy, like in the case of the NFIP which specifies how flood plans are required to be zoned for insurance purposes.

Urban flood control requires that corridors be reserved for passage of large flood flows, where appropriate. The locations for such corridors will normally be determined by tributary catchment areas so that if they are large enough to create major flooding, the corridors will be able to handle it.

Demand for flood control services does not involve water resource use but use of water facilities, especially when reservoir capacity is used for flood control when it could be used instead to increase the water supply yield of a stream. In this situation, flood control is considered a water management purpose, but the infrastructure is to be managed and not the use of water resources. Demands for flood control from such reservoirs address how the facilities are operated.

Environmental Water and Instream Flows

The requirements for environmental water extend to both quantity and timing and are controlled by water pollution stream standards and needs of plants, fish, and terrestrial animals that depend on water. Pollution standards are set of regulations, but requirements for water to nourish fish and wildlife and maintain healthy stocks of vegetation and biota in streams for natural systems are highly variable and dependent on multiple factors.

Environmental water shapes streams through annual flood cycles and maintains habitat in many ways. Naturally occurring instream flows support habitat at the border between aquatic and terrestrial ecosystems, such as on flood plains. Environmental water is also essential for aesthetic qualities of water.

In contrast to diversions for irrigation and municipal uses, quantifying environmental uses is more problematic. Even scientists may disagree on quantities of water required to support fish populations. A typical environmental study will involve specialists in different fish, bird, and animal species, as well as experts in hydrology, hydraulics, and plant life. By studying water needs from all views, a balanced view of water needs will emerge.

Unless water is provided for essential environmental uses, the impacts will be drier habitats, diminished vegetation stocks, and lower stream flows. These impacts are recognized by environmental impact studies, which identify the need to sustain environmental uses of water. The concept of ecosystem services offers a formal way to justify the use of water for natural systems when there are other demands for it.

Instream flows are waters flowing in rivers and smaller streams and remaining after all withdrawals, discharges and losses occur. Instream flows are the common responsibility of the players, including utilities, dam owners, regulators, power companies, irrigation companies, local governments, and other water users. Their quality and quantity are determined by dam releases, coordinated decisions, and unstructured actions. In addition to water quality and environmental flows, they include:

- Hydropower – flows to generate hydroelectric power
- Navigation – minimum depths for commercial navigation
- Recreation – flows for swimming, boating, fishing, and esthetics
- Water conveyance – carry water to points of withdrawal and use

Demands for instream water are normally set by regulations or by agreements. Some states have instream flow laws to establish them. Demands to serve hydropower, navigation, and recreation occur simultaneously with those for pollution control, environmental water, and use of channel capacity for flood control. A demand might be set by some form of agreement to supply, say, a certain flow for energy generation. Navigation might be similar in that an agreement to provide a depth of water might determine the demand. These are situation-specific and often require balancing based on shared interests to pursue economic goals (electric energy and navigation), social goals (water quality and recreation) and environmental goals (water for plants, fish, and wildlife). Demand for water-based recreation requires clean, abundant, and scenic water in streams, lakes, and reservoirs for boating, swimming, rafting, fishing, water skiing, picnicking, and sightseeing. Rivers are attractive to communities for creation of water-based developments. They offer possibilities for amenity-based developments such as bike trails. Flood plains are environmentally-attractive venues for mixed natural area uses in urban areas.

Questions

1   Explain the differences between water withdrawal use, consumptive use, and instream flow use.

2   In terms of diversion from streams, which water sector uses the most water? Which is the biggest user in terms of consumptive use? How might the differences compare in dry and humid regions?

3   What is the general trend of industrial water use in the United States? What is driving the trend?

4   What kinds of industries are more likely to use self-supply and what kinds are more likely to tap into water from a distribution system?

5   Why would high water-using industries find it attractive to locate in a country or state with weak enforcement of water and environmental laws?

6   What are the main drivers that determine the daily water demands in a city? Explain how you would use data about these driving variables to develop a model to estimate future water use in a city.

7   How are daily water demands used to determine the needed capacity of a water treatment plant?

8   Estimate annual average per capita water use in these US city types with: (i) mostly residential, small lots, humid region; (ii) mostly residential, large lots, arid region; (iii) some residential, mixed size lots, also with substantial commercial, industrial water uses, humid region?

9   Given how commercial water demands depend on type of activity and land use, what approach would you use to estimate future commercial water use in a given city?

10 Industrial water uses occur either by connections to distribution systems or by self-supply systems. Name two types of industries you might expect to use distribution systems and two that might be more likely to use self-supply.

11 Public water uses can be a substantial in cities. Make a list of five ways the public sector might use water and sort them from highest to lowest uses.

12 In many cities, leaky distribution systems cause major losses of water. How do these losses affect estimates of per capita demand when planning the capacity of systems?

13 How are requirements for the size of water distribution system piping determined?

14 It is generally understood that demand for water in a city is inelastic. What does this mean? Explain your answer.

15 In economics, the equilibrium level of producing a good for a market is set by demand and supply relationships. How does this principle apply or not apply to water conservation incentives in the case of water supply?

16 If demand for water decreases, what effects will be the effects on source of supply, utility revenue, water infrastructure, and operating costs?

17 How would demand for water supply in cities among low-income countries be different from demand in US cities? What are the factors causing the differences?

18 How is the capacity of a water treatment plant determined?

19 Compare the state of the art of water use forecasting for residential versus industrial uses.

20 Can USGS's estimate of average residential demand in the United States be used to plan a local water supply system? Why or why not?

21 How is demand for sizing the capacity of a wastewater collection system determined?

22 How are the requirements for capacity of wastewater treatment plants determined? Are they based on economic analysis or some other criteria? Explain.

23 In planning for operation of a combined sewer system, the frequency of allowable overflows is a main decision variable. Is selection of this frequency determined by economic analysis or how else might it be determined?

24 Policy leading to the CWA debated the feasibility of two approaches to setting standards: economic incentives by pollution pricing versus command-and-control by regulation. Which approach was selected and why? How does the CWA implement this approach?

25　How is the tradeoff between use of receiving waters for dilution and increasing capacity of wastewater treatment plants determined?

26　How does a stream standard under the CWA affect demand for water quality in natural systems?

27　Explain the concept of stream standards as compared to economic demand for wastewater treatment.

28　Explain why it is difficult to estimate the water requirements of ecosystems.

29　What are ecosystem services? Explain.

30　What categories of natural ecosystems have requirements for water?

31　How are requirements for natural ecosystem water determined?

32　If natural systems need certain quantities of water at certain times, how can this requirement be implemented in water resources management decisions?

33　If fisheries have cohorts of fish at different phases of life cycle, how can their water needs be determined?

34　How do stormwater quality requirements affect decisions about capacity and cost of systems in cities?

35　How are standards for capacity of minor stormwater systems determined?

36　If a stormwater system is designed to manage a two-year storm, in what sense is that a demand on the system?

37　How are standards for minor stormwater or urban drainage systems determined?

38　How and when did stormwater quality management become a requirement in cities?

39 Explain how the demand for flood control services is established in a general way.

40 For a typical situation where riparian properties along a stream are vulnerable to flood damage, are public decisions about risk management made in response to demand for flood control services? Explain.

41 Give an example of the demand for irrigation water as a function of cropping patterns.

42 Irrigation water is one of the inputs for crop production as a market activity that generates income for farmers. Can irrigation water be priced and allocated by a market system? Explain.

43 What factors explain how demand for irrigation water might vary along the scale from low-value to high-value crops? Explain.

44 The area to be irrigated by one surface water system flowing by gravity is called a command area. If some farmers decide to fallow their crops, will demand for water in the system fall on a proportional basis? Explain.

45 Can agricultural water use efficiency be measured in the same way as urban water audits?

46 Demands for water for hydropower production often compete with other needs. Will these demands be met by pricing on a supply–demand basis, or how would decisions be made?

47 Water-based navigation for commercial purposes occurs in large rivers and lakes. Which water management parameter(s) are critical in establishing the demand for water?

48 Demands for water-based recreation using streams and lakes can be substantial. How can these demands be valued to determine how they compete with other needs for water?

49 Name four uses of water as instream flows.

50 What are the institutional arrangements to work out coordination among these diverse interests and off-stream water users?

Problems

Urban Demand

A city with a population of 200 000 and mixed residential, commercial, and light industrial land uses has a per capita water demand of 150 gpcd. This is calculated as total water production from the treatment plant divided by the city's population. The city's population is forecast to be constant over the next 10 years but an increase in urban farming is expected to generate an increased water use of 5000 AF/year. What will be the total demand in gpcd after this new farming demand is added?

Population Growth and Demand Forecasting

In a city with current population of 100 000, the estimated population and water use in gpcd for the last 10 years are as shown in the table. If population growth is now projected at 1% per year for the next 20 years and if per capita water demand can be reduced on a linear basis by conservation rates and improved management by 20% over the 20-year period, what is your projection of city water demand in AF at the end of year 20?

Year	Population	gpcd
1	91 500	149
2	92 000	147
3	93 700	145
4	94 200	145
5	95 000	140
6	96 100	139
7	97 500	135
8	98 000	134
9	99 200	131
10	100 000	130

Demand for Water Treatment

A city of 100 000 people has an average annual water use of 175 gpcd. If the peak factor for designing the water treatment plant is 5.0 (ratio of peak hour to average hour for the year), what capacity is needed for the plant in million gallons per day so that it can meet demands during the peak hour?

Peak Demands

A city's peak day and peak hour demand will be important in designing storage and treatment facilities. Factors for these vary, but if peak day ratio is 2.0 and the peak hour ratio is 5.0 times the average hour, what will be the peak day and peak hour demands when average demand is 7.5 mgd?

Water Conservation

Monthly treated water demand in gpcd in a city of 100 000 is shown in the below table. If a conservation program can reduce demand by 25% in the outdoor watering months of May through September, what percentage savings in annual total water demand will result?

Jan	Feb	Mar	Apr	May	Jun	Jul	Aug	Sep	Oct	Nov	Dec
70	70	70	90	110	140	160	170	150	100	80	70

Conservation Rate

A family with three members left on a one-month vacation last July and forgot that a garden hose was running. By mistake, the hose discharged 5 gpm all month. The local water department allows homes to use 100 gpcd at a flat rate of $50/month. For any water use above that quota, it charges homes $8.00/1000 gal. What will be the total water bill for the month and how much of it will be an extra charge due to the hose what was left running?

Crop Water Demand

Estimate the evapotranspiration for a 100-acre field of corn with a growing season from 1 April through 30 September. The reference ETO values and crop coefficients are given in the table.

	ETO	Kc
January	1.24	
February	2.24	
March	3.72	
April	5.7	0.3

(*Continued*)

	ETO	Kc
May	7.44	0.6
June	8.1	0.9
July	8.68	1
August	7.75	0.9
September	5.7	0.7
October	4.03	0.3
November	2.1	
December	1.24	
Year	57.9	

Examples of Units

What is the water use in a house during January if there are three persons using an average of 125 gal each per day? (Water use = 3 * 125 * 31 = 11 625 gal = 11.6 TG = 15.5 CCF).

If a person's average use is 150 gpcd, what fraction of an acre-foot of water do they use?

(Use = 150 * 365/325850 = 0.168 AF).

At 150 gpcd, how many persons can an acre-foot supply? If the use was at 250 gpcd, as it could be in some western cities, how many persons can an acre-foot supply? (An acre-foot supplies 1/0.168 = 5.95 or 6.0 persons at 150 gpcd; at 250 gpcd, 1 AF will supply (150/250) * 5.95 = 3.6 persons).

A water treatment plant will deliver 15.0 million gallons/day at peak rate. If the daily peaking factor (peak day divided by the average day) is 1.5, how many homes with an average occupancy of three people can it serve with an annual average of 150 gpcd? (15 mgd/1.5 or 10 mgd. 10 mgd/150 gpcd = 66 667 people or 22 222 homes.)

For a town of 10 000 people and an annual average water use of 150 gpcd, how many acre-feet do they use in a year? Million meters^3? (10 000 people * 150 gal/person/day * 365 days = 547.5 MG = 1680 AF = 2.07 MCM.)

Daily consumption in a home is 140 gpcd. How many liters/day is that? (140 gal * 3.7854 l/gal = 530 l/day/capita.)

A water meter indicates that during the last 30 day month the customer has used 1925 units of 100 feet^3 of water. How many 1000 gal units is that? If three people live in that house, what was the consumption in gcpd? (For three people, per capita use is 160 gpcd.)

What is the average rate of water supply required for a city of 50 000 population, with a per capita use of 150 gpcd? (Water needed = 150 * 50000 = 7 500 000 gal/day = 7.5 mgd.)

The unit mgd is commonly used to specify treatment plant capacity but a city might use acre-feet to specify its annual need for a water supply. Based on the average need of 7.5 mgd, what is the average demand in acre-feet per year? (Water supply = 7.5 mgd = 8400 AF/year.)

5

Hydrologic Principles for Water Management

Hydrology and Water Management

This chapter explains the basic hydrologic principles needed to support decision making in water management. It draws from the broad field of hydrology to explain in a basic and practical way how its principles are applied across diverse problem sets. It is the first of three chapters focused on hydrologic topics and is followed by one to explain how water balances are used in management and another about how extreme value hydrology works in the case of flooding. Extreme value hydrology for low flows and drought is explained in this chapter, along with concepts of natural systems, water quality, and ecohydrology. The common element among these is descriptive hydrology to explain how the systems work.

The three chapters that are focused on hydrology explain basic concepts and computations, but they omit details that are readily available in many textbooks, handbooks, and government websites. While the discussion is at a basic level and not equivalent to a course in hydrology, the topics are presented in sufficient detail so a water manager can understand concepts and computations, while leaving specialized topics to other sources.

The main subjects of physical hydrology are atmospheric dynamics, precipitation, runoff, losses, and flows through streams, aquifers, and connecting paths. Water resources management situations may involve average or extreme conditions of these variables, and these often involve extreme wet periods when flooding occurs or dry periods that involve low flows and drought conditions. The deficits, shortfalls, and scarcity these create can occur by natural causes like meteorological drought or human causes like poor planning, infrastructure deficiencies, and waste.

Water Resources Management: Principles, Methods, and Tools, First Edition. Neil Grigg.

Water quality hydrology is useful to address runoff impacts on streams and aquifers. It provides a basis to assess chemical and biological aspects of the hydrologic cycle. The concept of biogeochemical cycling explains how biological, geological, and chemical substances like nitrogen, phosphorous, calcium, and carbon move along with the water cycle through the biosphere, lithosphere, atmosphere, and hydrosphere. Ecohydrology explains how water interacts with living things, especially plants and the biosphere. It is useful in problems like quantifying consumptive use of crops and other vegetation.

Physical Hydrology and the Water Cycle

The hydrologic cycle, or water cycle, provides the conceptual basis to analyze water management situations. As shown in Figure 5.1, the natural elements involved with the water cycle are:

- Atmospheric water and precipitation
- Evaporation and evapotranspiration
- Rainfall-runoff and losses in watersheds

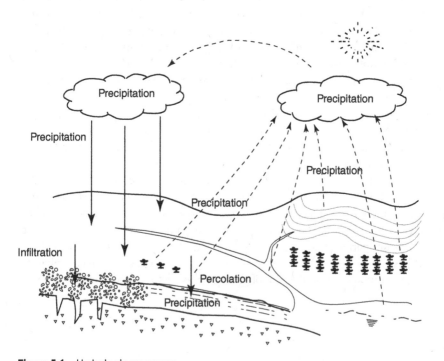

Figure 5.1 Hydrologic processes.

- Aquifers and ground water systems
- River flows, riverine networks, and sediment movement
- Floodplains and riparian areas
- Wetlands and small ponds
- Lakes and reservoirs
- Estuaries

Each of these hydrologic elements is critical in water supply, water quality, and the natural environment. Although water regulations and management strategies often address them separately, they comprise a unified natural water system. Ecologists recognize them as a unified whole and water management goals, whether supply, quality, or environmental in focus, depend on addressing them systemically. Debate over a regulation called the "Waters of the United States" (WOTUS) addresses the unification of these elements (see Chapter 20).

The figure is adapted from the cover of the NRCS National Engineering Handbook, which contains a gold mine of hydrologic information. It was redrawn and has some differences with the NRCS diagram, but the concept is like many other depictions of the hydrologic cycle. The USGS publishes attractive colored versions of the water cycle, and these can be easily located on its website.

Hydrometeorology and Climate

The atmospheric part of the water cycle interacts with land to create surface and subsurface water flows and storage. It is addressed by hydrometeorology, which blends meteorology and hydrology to explain the transfer of water and energy between land and the lower atmosphere. In turn, this explains how water yields, drought, and floods occur, along with long-term (climatic) and short-term (weather) events.

Energy from the sun evaporates water from land and water bodies to create water vapor, which is cooled to condense and become precipitation. Most atmospheric water evaporates from water surfaces and is transpired from plants (about 10%). All atmospheric water would cover the whole Earth with about 1 inch of water, uniformly distributed.

Climate is dynamic, and climate change occurs continually. If climate stayed constant in a particular region, the state would be referred to as stationarity. If you use historical data to analyze statistics of hydrologic phenomenon like rainfall and runoff, you normally would assume stationarity, which means a constant record over time, but in the long term, this assumption can be called into question. Climate change leads to non-stationarity, which must be considered in many water resources management situations.

Climate change is an ongoing phenomenon, which differs from place to place. It refers to the long-term change in average weather patterns, which are apparently caused by human activities, especially burning of fossil fuels that increases greenhouse gas levels and the Earth's average temperature. Some natural processes also contribute, including ocean warming and cooling like El Niño, La Niña, and the Pacific Decadal Oscillation and forces like volcanic activity and changes in solar energy. Impacts being observed include land and ocean temperature increases, rising sea levels, melting of ice in polar and mountain regions, and changes in extreme weather like hurricanes, droughts, and floods.

Precipitation that determines runoff during average and extreme conditions comes mainly as rainfall or snow. Rainfall at a point is normally measured by gages and the measured data are recorded, usually by a government agency. In the United States, the National Weather Service is the main agency that studies weather and makes forecasts. It is located within NOAA, which has centralized its environmental data in its National Centers for Environmental Information (https://www.ncei.noaa.gov). The website provides a portal to order many types of environmental data. States have climate offices that may provide more localized data. They can be identified through the American Association of State Climatologists (https://stateclimate.org).

If you need data for a study that required analysis of 100 years of point rainfall data at a given station, say Denver, Colorado, you could obtain it immediately from https://www.weather.gov. You would select Colorado from the map offered, then Denver and "Climate and Past Weather." You could then enter the date range and obtain the data. For example, obtaining the 100-year record for Denver from 1921 to 2020 yielded this result of mean values for precipitation in inches per month:

Jan	Feb	Mar	Apr	May	Jun	Jul	Aug	Sep	Oct	Nov	Dec	Year
0.47	0.56	1.17	1.78	2.29	1.61	1.77	1.49	1.09	1.02	0.76	0.51	14.49

Denver receives snowfall so some of the totals reflect snow water equivalent. From the times series of these data, you could conduct various analyses to study risk, variation, or other parameters of interest.

For flood or stormwater studies, rainfall intensity–duration–frequency (IDF) curves are needed rather than point totals. The NWS Technical Paper 40 was the original standard reference with maps and areawide statistics. It was created by NOAA in 1961 from available data of the time. It provided return periods from one to 100-years and rainfall durations from 30 minutes to 4 days. It has been replaced by Atlas 14, which is being updated to address climate change. https://hdsc.nws.noaa.gov/hdsc/pfds. The data portal for Atlas 14 enables you to access IDF estimates at a station. For example, for Montgomery, Alabama, at the airport the data show for a 30-minute duration:

Frequency, years	1	2	5	10	25	50	100	200	500	1000
Depth, inches	1.17	1.34	1.62	1.84	2.14	2.37	2.59	2.81	3.10	3.31

The 90% confidence levels are also provided with the data, which data can be graphed to provide IDF curves.

The contribution of snowmelt to water supply and flooding is studied through snow hydrology. In some US western states, like Colorado, snowmelt provides a large share of the water supply. Snowpack is influenced by location in watersheds and vary due to solar energy, vegetation, and winds. Variables include snow depth, snow water equivalent, and snow density. Snow water equivalent is the most useful parameter for water supply studies as it is the depth of water available in an area if all snow melted. Snowmelt rates that influence flooding depend on several factors and are hard to predict. The main agency reporting snow water equivalent is the Natural Resources Conservation Service (NRCS). Data are available at: https://www.nrcs.usda.gov/wps/portal/wcc/home/snowClimateMonitoring/snowpack/snowpackMaps

Surface Water Hydrology

When assembling supply portfolios, water resources managers usually choose from available natural water sources like surface water in streams, stored water in reservoirs and lakes, and groundwater. Some managed sources may also be used, such as rainwater and stormwater harvesting, aquifer storage-recovery systems, reclaimed water, and desalted water.

Although many water supply sources are from groundwater, there is more emphasis on surface water hydrology. Many reference sources are available, such as the *NRCS National Engineering Handbook* and technical manual for the computer program HMS (see Chapter 9).

When analyzing surface water sources, the goal is usually to compute runoff, either under average or extreme conditions. It is computed as precipitation less losses from evaporation and transpiration, infiltration, and interception. Temperature, wind, and vegetation types affect evaporation and transpiration, and antecedent conditions determine the capacity of soils to absorb and infiltrate water. Sublimation losses that occur from ice or snow may be involved in some cases.

The runoff that results from precipitation after losses and baseflows from groundwater are counted determine the magnitudes of flows in stream channels and floodplain zones. Losses are difficult to estimate with accuracy and baseflow estimation is hampered by difficulty of knowing aquifer conditions.

Figure 5.1 shows the main hydrologic processes that determine runoff. Losses shown include evaporation and transpiration, infiltration, and interception. Recharge, runoff, interflow, and baseflow are shown to illustrate the mobility of water after it falls as precipitation.

Evapotranspiration (ET) is a measure of total water consumption over an area from surface evaporation, soil moisture evaporation, and plant transpiration. Water surface evaporation data normally report pan evaporation. Evaporation from a water body is measured by lake evaporation, which is lower than pan evaporation due to the complexity of lakes. Lake evaporation is normally about 0.70–0.75 of pan evaporation.

Average areawide evaporation can be estimated from maps. NOAA has published Technical Report NWS 33, Evaporation Atlas for the Contiguous 48 United States. Also, evaporation data and model estimates are available from NOAA's Climate Prediction Center. Local conditions vary, and pan evaporation data from local areas may be required for specific studies.

Computation of ET is required in any situation where water consumption must be estimated, like water rights studies. Computation of ET is discussed in Chapter 3 (demand). Several formulas are available. The basic approach to calculate crop ET is by multiplying a reference evapotranspiration, ET_0, by a crop coefficient, K_c.

Infiltration is an important parameter in surface hydrology, especially for storm runoff calculations. It depends on soil type, but also on antecedent conditions, vegetation, slope, and other variables. Infiltration can be measured in the field or estimated using one of the available equations. For example, one popular method is based on the Horton Equation. It considers that when soil is dry, the infiltration rate will be greater than after the soil become saturated. So, the infiltration rate decreases during a storm. Horton's equation is:

$$f_t = f_c + (f_o - f_c)^* e^{-kt}$$

where f_t, f_c, and f_o are infiltration rates at time t, at equilibrium, and initially, and e^{-kt} is a rate factor with k being a soil-specific decay constant.

Interception refers to water that adheres to surfaces like leaves and eventually evaporates. Depression storage means water captured in small holes and uneven surfaces of the land. They can be estimated, but they are so variable that estimates are difficult.

After losses are deducted from precipitation, runoff enters stream channels from the upper reaches of watersheds to outlet points. Small watersheds emit discharges to small channels that join others to create larger streams, and this process continues until large rivers discharge into the seas. This creates a hierarchy of stream channel sizes, which can be charted through a concept called "stream order," which is widely used in geomorphology and river mechanics.

The main channel, which is called the "floodway" for flood insurance purposes, lies in a valley and is bordered by floodplains, which have been created by geologic forces. The riparian corridor comprises the channel and floodplain wetlands that sustain the aquatic ecosystem. Maintaining it in healthy condition is critical to the functioning of natural systems. This requires adequate instream flows, which are critical to sustainability of ecosystem integrity.

Data for runoff in the form of stream flows can be obtained from the USGS and runoff models will utilize these data for calibration and verification (see Chapter 9). Analysis of stream flow records through study of time series is important in studies of water supply yield, which is explained later in this chapter.

Lakes, reservoirs, ponds, and wetlands are important components of stream systems. Lakes can be natural or created by dams to form reservoirs. Reservoirs for water storage are important constructed elements in water systems because they determine the volumes of flow available for release to streams. The way than a reservoir works is explained in Chapter 7. Smaller ponds and wetlands are widely distributed across watersheds and are especially important to sustain ecosystems. The connections among small and large streams, ponds, and wetlands are important to the hydrologic cycle and to biodiversity. They are addressed through a regulatory program known as the WOTUS (see Chapter 20).

An estuary is the mass of water formed by the confluence of a freshwater channel and the sea. Estuaries are important end points in many natural water systems because of their biological productivity. Managing estuaries is an important scenario in water resources management and is discussed in Chapter 15.

Low Flow Hydrology

Water supplies from surface water or groundwater sources can become depleted due to lack of precipitation and decrease in base flows. If the capacity of a water supply source to meet demands is marginal, it is likely to fail during low flow events or any period of decrease in water availability. Streamflow varies statistically and drought patterns can be persistent. If the surface supply is a lake, its water level can fall. In both cases, regulatory controls on withdrawals may cause depletion of supply. Groundwater levels can also fall, and the cost of pumping may rise due to greater head to pump against. Groundwater withdrawals can also be limited by regulators.

The deficits, shortfalls, and scarcity created by low flows can occur by natural causes like meteorological drought or human causes like poor planning, infrastructure deficiencies, or waste of water. To improve any of these situations requires risk analysis that considers the ratio of supply to demands to study water security under drought conditions. Having access to water supply reserves is a hedge against failure and requires a margin of safety. Water supply agencies should be self-reliant while working with others toward collective security. In many cases, more water storage and management capability may be needed.

The relationship between expected supply and demand determines the risk of water supply failure and security of adequate supplies. The analysis of surface water, groundwater, and reservoir sources is somewhat different. Also, determining water supply security for systems involving multiple sources involves different approaches.

The adequacy of water supplies is measured by the probability that the system will run out of water. This is usually presented as the return period of droughts of certain levels or as the annual probability of running short. While the general concepts of failure probability and return period are well known, they are complicated for droughts because duration, magnitude, and starting and ending times are unknown.

Rivers and streams comprise the source of water for many systems. Diversions have important effects on the stream systems and must be limited to the sustainable carrying capacity of the stream. This capacity is measured by water yield, which is a statistical quantity because of the time-varying discharge.

Generally, yield means the average supply available for some time period; for example, an average yield of 50 000 AF (61.7 MCM) for a certain river. Safe yield is determined statistically and is the water produced over some averaging period (like a day, week, month, or year) in a given time span, such as 50 or 100 years. The concept of safe yield has similar terms in use, such as dependable yield, firm yield, and reliable yield. It also has a parallel concept called firm power in electricity production.

For surface supplies, safe yields are estimated by statistical studies in situations such as withdrawal from streams, from reservoirs, and from water supply systems consisting of streams and reservoirs. For ground water, safe yield is the amount of water that can be withdrawn safely from an aquifer without overdraft, mining, contamination, or other failure.

For a stream diversion point, safe yield will be the lowest time-averaged flow for the time period under consideration. For example, if it is desired to know the yield of a supply for B, which is diverted at A, then the low-flow statistics at A must be evaluated.

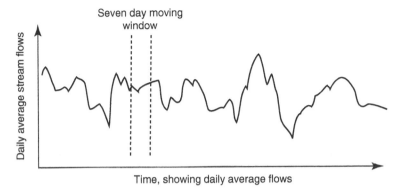

Figure 5.2 A hydrologic time series showing weekly averaging.

An example would be the 50-year weekly safe yield, or the lowest weekly average flow in a period of 50 years. This is illustrated by Figure 5.2, which shows a time series of flows averaged for successive weeks.

The computational procedure begins with a determination of plotting positions for weekly mean flows. All weekly flows in the data record are tabulated and sorted from highest to lowest. The plotting position, which is a measure of probability, is computed as:

$$PP = 100^* (m / (n+1))$$

where m is the ranking of the weekly flow and n is the total number of weekly flows.

The data can be plotted on probability paper or used in a probability distribution, such as the log-Pearson Type III that has been recommended for use by an Interagency Advisory Committee Water on Data. Detailed procedures are available at (Risley et al. 2008).

For a single reservoir, safe yield for a period under consideration will be the amount that can be withdrawn by utilizing the storage available without exceeding reservoir limits. The computation will be based on a water balance, which is explained in the next chapter.

Water Quality Hydrology

As water flows through its cycle, its quality changes due to natural causes and human activity, or anthropological effects. The changes can include chemical, biological, and physical effects due to phenomena such as erosion and sedimentation, leaching of earth metals and other inorganic compounds, and animal wastes. The main problems stem from human effects, especially urbanization and agriculture.

Many policy studies of water quality have been completed, and they usually explain that water pollution is caused by land use patterns, transportation systems, housing, urban centers, and energy consumption. Land development results in runoff of water, sediment and chemicals and can cause wetland and habitat destruction. Urban runoff causes contamination through release of sediment, organics, oil, and toxic chemicals. Wastewater systems are point sources of pollution. While treatment plants remove contaminants, they may fail, may not remove toxics, and may be bypassed by combined sewer overflows. Transportation systems involving roads, airlines, ships, and pipelines are major sources of contamination and they can destroy land or aquatic habitat, like through cutting of wildlife corridors and dredging.

Agricultural activity discharges sediment, nutrients, and chemicals in runoff, and can destroy wetlands and riparian environments. Runoff from animal production sends nutrients to streams and lakes, causing eutrophication. Deposition from the atmosphere may also impact lakes and estuaries. Agricultural chemicals can threaten groundwater. Industries such as manufacturers, power generators, waste treatment companies, and mines discharge conventional and toxic substances and thermal pollution into streams. Legacy problems such as acid mine drainage, polluted groundwater, and contaminated cause water quality issues.

Water quality assessment can provide essential information about levels of oxygen, bacteria, chemicals, and nutrients. Many parameters are involved and are often difficult to measure. In addition to data, water quality modeling is used to support water quality management (see Chapter 15).

Aquifers and Ground Water Systems

Hydrogeology is important because many water supply systems utilize groundwater, and it is important to protect it and manage it sustainably. Water managers might be required to analyze an aquifer, manage it to develop a sustainable supply, and protect the aquifer from contamination. For these, it is important to understand aquifer properties, movement of groundwater, how the groundwater connects to surface water, chemical changes, use of wells and recharge facilities, and controls on the use of groundwater.

Groundwater occurs in different types of aquifers (Figure 5.3). Alluvial aquifers are adjacent to streams and the connection of ground and surface water comprises a stream-aquifer system. The groundwater just beneath and adjacent to the stream is called the hyporheic zone. Alluvial aquifers are unconfined, which also

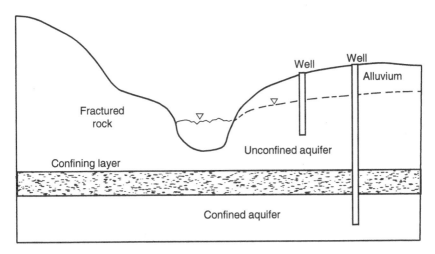

Figure 5.3 Types of aquifers.

describes other aquifers without overlaying layers of rock. When aquifers are beneath layers of rock, they are considered as confined. When these aquifers are under pressure, they can become artesian or flow due to the pressure. If an aquifer clearly discharges to a stream, it is considered as a tributary aquifer. Springs are small scale examples of water from tributary aquifers. Conversely, a non-tributary aquifer lacks a clear connection with a stream. A non-tributary aquifer may contain water that has been in place for so long that it is fossil water.

Aquifers are porous media and can contain large quantities of water available for use if they are managed properly. Storage of water is measured by the porosity, or the fraction of the total pore volume within the total volume of the aquifer. Closely related is the specific yield, which measures the relative volume of water that can be extracted by pumping. The water available in an aquifer is thus the area times saturated thickness times specific yield (as a fraction).

Storage in aquifers ranges from such large quantities down to very little, as in the aquifers present in rock cavities in mountain regions. For a large aquifer, the USGS estimated that in 1980 the Ogallala aquifer in the Central Plains held about 3.25 billion acre-feet of water (4009 BCM). That is about 125 times the 26.1 MAF stored in Lake Mead when full. Such large quantities of water offer many possibilities for productive use when managed well. http://plainshumanities.unl.edu/encyclopedia/doc/egp.wat.018.

Groundwater moves slower than surface water and is exposed to different chemical and biological environments. It infiltrates from the surface, moves though the unsaturated zone, and travels through aquifers at rates that depend on their permeability. An aquifer with large sediment grain sizes, such as sand and gravel, will have much higher travel times than one with smaller grain sizes, such as clay materials.

The aquifer property that measures water mobility is hydraulic conductivity, or capacity of the aquifer to allow flow of groundwater. The general equation to measure it is Darcy's Law, which can be written in a simple form as:

$$Q = K^* A^* S.$$

where K is the coefficient of hydraulic conductivity, a property of the aquifer; A is the cross sectional area; and S is the gradient.

As groundwater moves, it is subjected to geochemical influences that alter its quality. Earth elements can dissolve and occur in groundwater in varied concentrations. Deep groundwater that has been in contact with rocks for a long time may have greater concentrations of chemicals than shallow groundwater.

Dissolved components include compounds of calcium, sodium, magnesium, and various other compounds and metals. All dissolved constituents in water are called total dissolved solids or TDS. Trace elements also occur, such as silicon and iron. The pH of water is important because acidic waters can damage equipment and alkaline waters may cause precipitation and scaling, such as to clog up a well or pump.

Carbonates react readily with water are present in many rocks, so carbonate chemistry is important in groundwater analysis. Groundwater controlled by carbonate reactions will have relatively high calcium and may be hard or damaging to pipes and other equipment.

Arsenic in groundwater is a threat to health and is a worldwide issue. It is particularly threatening along floodplains in Bangladesh where millions of people use groundwater from small tube wells. It was discovered that many of the wells have high arsenic levels and are at risk of poisoning that leads to health effects like cancer, diabetes, and other diseases.

Wells are the infrastructure of ground water and need to be managed carefully. The figure shows a shallow aquifer and well drawdown. In addition to wells for withdrawing water, there are also recharge wells to pump water back into aquifers. Aquifer recharge can also be by seepage from ponds.

For ground water, safe yield is the amount of water that can be withdrawn safely from an aquifer without overdraft, mining, contamination, or other failure. Safe yield for streams and reservoirs was discussed earlier, and the principle is the same, that is, not to expect more water than can be provided sustainably.

In a tributary aquifer, groundwater will move toward a stream and eventually flow into it. This creates stream-aquifer interdependence and might need to be regulated to avoid robbing water from the stream unknowingly. Stream-aquifer modeling is important to assess the degree to which groundwater use affects streams.

If groundwater overdraft occurs, there is a risk of subsidence and the lowering of ground elevations on a permanent basis. This has been a serious problem in many places with heavy drafting of groundwater. When groundwater levels fall in coastal areas, the risk of saltwater intrusion also increases.

Controls on the use of groundwater are discussed in Chapter 20. Protecting aquifers from contamination is also important. How transport of contaminants occurs in different aquifers requires analysis and modeling. Regulations to control aquifer contamination are discussed in Chapter 20.

Quantitative analysis of groundwater flow requires computer-based modeling. The basic relationship used for groundwater flow is Darcy's equation, which expresses flow as a function of head and is incorporated into models like MODFLOW, that is managed by USGS (see Chapter 9).

Ecohydrology

Ecohydrology explains interactions among ecological systems and the water cycle. Ecology's parent discipline is biology and is fundamental to understanding how natural water systems work. Natural ecosystems, or groups of plants, animals, and microbes in a common environment, depend on the water cycle for life. That is why drought is so devastating on ecosystems.

As both hydrology and ecology are broad subject areas, it means that ecohydrology is an open-ended area. Emphasis is on evapotranspiration of water from plant leaves and explanation of how plant roots draw water and nutrients to the stems and leaves and transpire some of it to the air. The rate of transpiration depends on numerous factors, such as temperature, humidity, sunlight wind, and others.

A short explanation of ecohydrology is provided here, and Chapter 8 explains how it is important to assess ecosystem services in production of food and water, control of climate and disease, supporting nutrient cycles and crop pollination, and in providing other societal benefits.

Water managers deal with terrestrial ecosystems (on the land) and aquatic ecosystems (water environments), and with their transition zones in watersheds, wetlands, or riparian strips. The water environment for aquatic ecosystems includes the natural water systems in streams, lakes, aquifers, lakes, estuaries, and oceans. In these, nutrient cycles provide essential chemical elements to plants and animals. Solar energy flows through light to organisms and their environments and by photosynthesis creates chemical energy in the form of plant food.

The watershed and stream reach are ecologic accounting units for aquatic and terrestrial ecology. The stream is a hydro-ecologic environment that integrates or cumulates all aspects of land use and water management. In the stream environment microorganisms feed on organic matter from runoff and deposition. If the balance is right, they will be healthy for fish and macroinvertebrates and the rest of the food chain.

As nursery habitat for fish, birds, and other wildlife wetlands protect ground water supplies, purify surface water by filtration and natural processes, control erosion, and provide storage and buffering for flood control. Estuaries and marine environments also sustain important ecological systems.

Reference

Risley, J., Stonewall, A., and Haluska, T. (2008). Estimating Flow-Duration and Low-Flow Frequency Statistics for Unregulated Streams in Oregon Scientific Investigations Report 2008–5126. https://pubs.usgs.gov/sir/2008/5126.

Questions

1 What kind of science is hydrology and what are its main topics? What is the field of hydrometeorology? How does it relate to hydrology?

2 Sketch the hydrologic cycle and indicate the principal natural water systems shown by it.

3 Explain base flows and how they influence instream flows and groundwater levels.

4 What is meant by global climate change and how is it expected to affect average and extreme rainfall?

5 Explain the concept of stationarity and how it relates to the changing climate.

6 What is stream order and how does it relate to watershed hydrology?

7 What is an estuary and why is it so important in the hydrologic cycle?

8 What is the physical cause of rainfall? Under what conditions might extreme rainfall occur?

9 What indicator is used to measure the water content of snowpack?

10 Name categories of precipitation data that would be used for water supply or flood control planning.

11 What is the main US agency that manages data for with climate and precipitation? What is the main reference publication to obtain general precipitation information in the United States? How do you obtain evaporation data?

12 What are antecedent conditions in hydrology and why are they important?

13 Explain the physical mechanism that causes evaporation from a water surface including the influences of temperature and wind.

14 Explain how you would estimate evaporation from a reservoir surface.

15 What is the difference between pan evaporation and lake evaporation?

16 Define evapotranspiration and explain the basic approach to computing it.

17 Explain what causes infiltration and how different types of soils affect it from the beginning until the end of the storms.

18 Explain the factors that would drive seepage losses from a reservoir.

19 Explain how vegetation experiences transpiration and causes interception during rainfall.

20 What is the biogeochemical cycle? Explain how it relates to the hydrologic cycle.

21 Using your sketch of the hydrologic cycle, indicate how water quality changes due to natural and human causes. Include sediment, biota, bacteria, and leaching of materials.

22 How might a reservoir be operated to enhance water quality downstream?

23 Sketch a groundwater system to show unconfined, confined, and tributary aquifers, a well withdrawing water from a shallow aquifer, and drawdown when the well is discharging.

24 Explain how to compute the water available to withdraw from an aquifer.

25 Sketch a stream-aquifer system and indicate the flow of water in it.

26 Write the Darcy Equation for groundwater and define the variables.

27 Explain how an aquifer storage recovery system works, what it might achieve, and what kind of concerns it might create.

28 Identify the influences on groundwater that affect the chemistry of surface water.

29 How does the legal issue of waters of the United States (WOTUS) relate to the hydrologic cycle?

30 Sketch a stream cross-section and identify the main channel and flood plains. Indicate on the sketch the approximate levels of average flow and extreme flows, like the 100 year flood.

31 How does non-stationarity affect the statistics of low stream flows?

32 Show how you would compute the 7 day-low flow on a stream for 10 and 100 year return periods.

33 What is a hydrologic time series? Give an example as it might be used in an analysis of safe yields.

34 Would you use a normal distribution or an extreme value distribution to analyze statistics of low stream flows? Explain.

35 In flood statistics, an extreme flow is an "event." Why are statistics of a drought event more complex than those of flooding?

36 What is paleohydrology? Give examples of how it can be used to analyze hydrologic phenomena. How are tree ring studies useful to study drought?

37 What are the basic causes of drought?

38 Define drought. What is the difference between a "meteorological drought" and a "hydrologic drought"?

39 What is meant by the statement drought is a "creeping disaster," and what would be a "sudden disaster."

40 Explain how drought intensity is measured and how its return period is measured, as compared to flood.

41 Formulate an example of a simple drought index and how it might be used in water management.

42 In the United States, how is the extent of regional drought intensity displayed to the public?

43 Define safe yield of a reservoir and explain how a reservoir increases the safe yield of a stream.

44 How does the concept of an instream flow requirement relate to the safe yield of a stream diversion?

45 Explain what is meant by a 50-year monthly water yield of 1000 cfs on a stream.

46 Define safe yield of surface water sources, wells, and a system of multiple water supply sources.

47 What is the difference between a reservoir's storage capacity and its yield?

48 Explain how you would compute the safe yield of two streams which are independent and you have the time series of each one.

49 Define ecosystem services and give an example of how natural water systems provide them. Give an example of how a natural water system could replace a built water system while providing ecosystem services.

Problems

Point and Area Rainfall Depths

During a storm, point rainfall depths will vary across a watershed (Figure 5.4). Methods to estimate total precipitation from point totals are based on interpolation and a summation of values. In the isohyetal method you draw contour lines of equal rainfall, compute the subareas between them, and totals for these subareas. For the watershed and point rainfall values (in inches) shown, use the

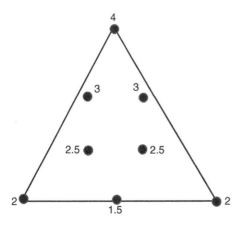

Figure 5.4 Rain gages in a triangular region.

isohyetal method to estimate the average depth of precipitation over the total area. The area shown is an equilateral triangle with base = 1 mile. The interior rain gages are spaced at 1/3 intervals across the base and along the height as shown.

Average Precipitation and Evaporation

Average monthly precipitation and temperature for Denver, Colorado, and New Orleans, Louisiana, are shown in the tables below. New Orleans is located on the Gulf Coast and Denver is a midcontinent state. Given that evapotranspiration is generally a function of temperature and humidity, the temperatures shown are approximate measures. Analyze the data and summarize your opinions as to: (i) the climatic influences that are determining the monthly fluctuations in precipitation for the two cities, and (ii) the likelihood of significant or runoff as determined by the excess of precipitation over hydrologic losses.

Denver

	Jan	Feb	Mar	Apr	May	Jun	Jul	Aug	Sep	Oct	Nov	Dec	Year
Precip	0.47	0.56	1.17	1.78	2.29	1.61	1.77	1.49	1.09	1.02	0.76	0.51	14.49
Temp	30.7	33.1	39.8	47.7	57.2	67.6	74.0	72.1	63.5	51.4	39.2	31.5	50.4

New Orleans

	Jan	Feb	Mar	Apr	May	Jun	Jul	Aug	Sep	Oct	Nov	Dec	Year
Precip	5.18	4.13	4.36	5.22	5.64	7.62	6.79	6.91	5.11	3.70	3.87	4.82	63.35
Temp	52.0	55.9	62.0	68.0	75.5	81.0	82.9	82.8	78.8	69.5	59.4	53.8	68.5

Precipitation Frequency Data

Use the Internet to find the National Weather Service's Precipitation Frequency Data Server, https://hdsc.nws.noaa.gov/hdsc/pfds. If the url does not work, search for the data server by its name. Go to North Carolina, then select the Raleigh–Durham Airport. Find the rainfall-frequency data for a 60-minute duration.

Intensity-Duration-Frequency Curves

For the airport at Montgomery, Alabama, the data in the table show rainfall depths for a 30-minute duration. Assume that the depths for 15 minutes will be 1.5 times these values and for 1 hour will be 0.75 times these values. Sketch the rainfall intensity–duration–frequency curves for the three rainfall durations.

Frequency, years	1	2	5	10	25	50	100	200	500	1000
Depth, inches	1.17	1.34	1.62	1.84	2.14	2.37	2.59	2.81	3.10	3.31

Infiltration

When a rainfall event starts, the initial infiltration rate is 4 inches/hour. After 12 hours it is constant at 1 inch/hour. Using the Horton equation shown, compute the infiltration rate at four hours into the storm. The value of k is 0.5/hour.

$$f_t = f_c + (f_o - f_c)^* e^{-kt}$$

where f_t, f_c, and f_o are infiltration rates at time t, at equilibrium, and initially, and e^{-kt} is a rate factor with k being a soil-specific decay constant.

Evaporation

If a supply canal is 20 miles long and has a surface width of 40 feet, what would you estimate the evaporation loss in acre-feet to be for June, July, and August? A nearby reservoir shows the following pan evaporation rates in inches for those three months: 17.38; 16.05; and 14.69.

Channel Flow

Uniform flow in open channels (Figure 5.5) is usually analyzed using the Manning Equation, which is:

$$Q = (1.49 / n)^* A^* R^{2/3*} S^{1/2},$$

where Q is the discharge in cfs; n is a roughness factor; A is the cross-sectional area; and S is the channel slope downstream.

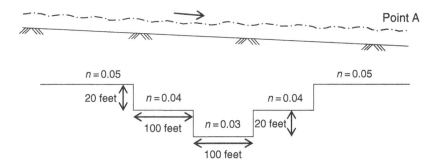

Figure 5.5 Channel cross section and profile.

The river reach shown has a uniform slope of 0.00075 feet/feet (see the profile on top) and a uniform cross section (shown at the bottom) throughout the reach.

1) For a stream flow of 25 000 cfs compute the depth of flow. Make any assumptions you deem necessary and state them.
2) Now, assume that a dam has been built at Point A. Explain the steps you would take to determine the new water surface profile for the same stream flow into the reservoir and as a function of the rate of release of water from the reservoir. You are not required to compute anything, but you could comment on how to model the change after dam construction.

Safe Yield from a Stream

Visit the USGS website (https://waterdata.usgs.gov/nwis/rt) for a stream of your choice and identify a time series of daily flow data (if the URL has changed, search for USGS streamflow data). Convert the daily flows to weekly data. Order the weekly data from lowest to highest (m = 1–n). Assign plotting positions ($P = m/n+1$). Plot the values on probability paper. Fit a probability distribution (like log-normal). Estimate low flows for 10, 50, and 100-year frequencies.

Safe Yield from Stream and Reservoir

During a five-year period, the following average annual flows enter a reservoir: 100, 200, 50, 150, and 400 cfs. The reservoir can hold 50 000 acre-feet, but its starting contents are 25 000 AF at the beginning of the five-year period. Assuming no losses, what is the approximate maximum constant discharge that can be delivered over the five years without ever running out of water in storage?

Water Available in an Aquifer

An unconfined aquifer with mostly sand and gravel has an estimated specific yield of 0.25. The average saturated thickness is 50 feet. How much water in AF might be extracted from pumping a portion of this aquifer with area = 1 mile^2?

Stream-Aquifer

An example of a stream-aquifer problem is the pumping well that affects streamflow (Figure 5.6). Determining the effects of the well on the stream requires a model, but you can make an order-of-magnitude estimate. For the situation shown, estimate the effects on the stream due to the pumping well. Assume that before pumping, the stream flow opposite the well is 5 cfs and the contribution of groundwater to the stream is 1 cfs/mile. The well pumps at 1000 gpm. As shown in the top view, the diameter of the well drawdown is one-half mile.

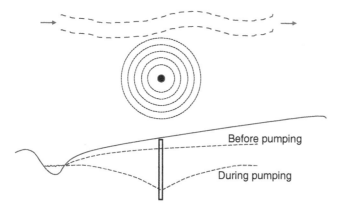

Figure 5.6   Pumping well by a stream.

Gaging Station

Average monthly flows for a gaging station are given below in cfs. Find the total flow volume for the year in acre-feet and in million cubic meters.

	Q, avg, cfs
Jan	112
Feb	187
Mar	375
Apr	500
May	650
Jun	575
Jul	387
Aug	305
Sep	261
Oct	185
Nov	150
Dec	112

6

Water Balances as Tools for Management

Introduction

Among the many concepts of hydrologic science, the basic water balance is perhaps the most useful tool for water analysis and management situations. Water balances are used in many scenarios, including water rights determination, permitting of water uses, allocation of stream-flows, and reservoir releases. They are used widely in regions of water scarcity and are also needed during periods of shortage in humid areas. Applications include small catchments through large river basins, irrigation command areas, political jurisdictions like cities, districts, states, and nations, and physical systems for water distribution and storage.

A water balance is an assessment of inflows, outflows, and losses during a time interval for an accounting unit such as a watershed. It is like a balance sheet in financial accounting, which is used for analysis and decision making and requires data on financial variables. It is a hydrologic concept because it requires data on basic variables such as inflow, outflow, losses, and diversions. Often, the values of the variables are uncertain, and the water balance yields only approximate results. Regardless, it remains the gold standard tool for many water management situations.

The water balance concepts discussed in this chapter can provide information relating to any accounting unit, the world, a state, a river basin, a small watershed, or an urban supply area, which can be the basis for assessing losses in distribution systems. Chapter 5 explained the hydrologic variables used in the water balance. This chapter outlines the basic concept and applications of the use of the water balance in management studies.

Water Resources Management: Principles, Methods, and Tools, First Edition. Neil Grigg.
© 2023 John Wiley & Sons, Inc. Published 2023 by John Wiley & Sons, Inc.

Concept of the Water Balance

The concept of the water balance stems from the basic equation of hydrology, which explains flows into and out of some system, like a reservoir or watershed:

$$I \times \Delta t - O \times \Delta t = \Delta S$$

where I is the rates of all inputs and O is the rates of all withdrawals and losses in the accounting unit during the time increment Δt. The variable ΔS has units of volume and is the change in storage that occurs in Δt.

Inputs to a system include precipitation, snowmelt, inflow from upstream and imported surface and groundwater (Figure 6.1). Losses include evaporation, consumption, evapotranspiration, and seepage. In urban settings, measurements of uses, losses, and unaccounted water are also required.

Water Balances in Watersheds and River Basins

The watershed or catchment is a common water management accounting unit. It can vary in size from the small watershed (even less than an acre) to a large river basin. There is no global standard to classify watershed sizes, but the United States has a system to classify them by area (US Natural Resources Conservation

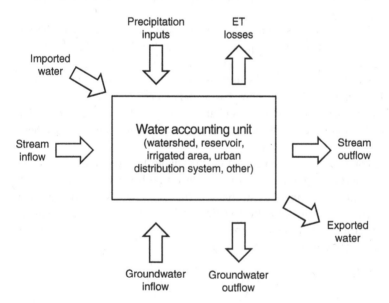

Figure 6.1 Inputs and outputs for a water balance.

Table 6.1 Classification of watersheds and river basins.

Name	Level	HUC[a] digits	Number in US	Average area, miles^2	Average area, km^2
Region	1	2	21	177 580	285 788
Subregion	2	4	222	16 800	27 037
Basin	3	6	352	10 596	17 052
Subbasin	4	8	2149	700	1126
Watershed	5	10	22 000	227	365
Subwatershed	6	12	160 000	40	64

[a] HUC = Hydrologic Unit Code.

Service 2022). Table 6.1 shows watersheds from an average of 40 mile^2 to regions covering several states.

There is a related concept of stream order, which aligns somewhat with watersheds by relative size but depends on geophysical properties. The concept of stream order assigns numbers to streams on the basis of number of tributaries, such that the smallest stream has order one and rivers have higher stream orders. Watershed sizes will not correspond exactly to stream order, but the related concepts are useful in mapping and analysis.

In any setting, small watersheds feed tributaries of larger streams. While there is no standard definition of a small watershed, the terms in Table 6.1 suggest that a 12-digit subwatershed is small, but much smaller watersheds require analysis in some applications. The 12-digit HUC with an average area of 40 mile^2 corresponds to the scale of a medium city of 5000 people/mile^2 and 200 000 people. An average county of 500 mile^2 would contain about 12 subwatersheds at the 12-digit HUC level and two at the 10-digit size. An example of the largest level would be the Missouri River Basin above St. Louis with more than 500 000 mile^2 in the region.

As a fundamental unit for hydrologic accounting, the watershed provides a basis for simulation models to support decision-making. A fine-grained model will include characteristics of very small watersheds because larger ones lack homogeneity. Even 12-digit HUC watersheds are sufficiently heterogeneous in land characteristics not to be treated as homogeneous hydrologic units for simulation modeling.

For local situations, the 12-digit HUC watersheds can be subdivided. For example, a 40 mile^2 watershed could be decomposed to 1 mile^2 units, and these could be further divided into smaller units. The Rational Method to compute stormwater runoff (discussed in Chapter 7) is usually limited to 200 acres (80 ha), so a 1 mile^2 watershed would include slightly more than three of these. At the very small scale, the drainage area of a city block, say 5 acres, could be served by a single drainage pipe and be the unit for a calculation of runoff. Even a roof area at around 1500 feet^2 (140 m^2) is tributary to drains and can be analyzed in a similar way.

Water Budgets for the World, Nations, States, and Cities

Large scale water budgets are useful for macro-level studies such as for climate change, where the global water budget has been studied extensively. Estimates at the global scale generally show that about 70% of the earth's surface is covered with water, some 97–98% of water is in the oceans, only 0.001% in the atmosphere, and the rest on the land (University of Illinois 2020). Further details show the relative percentages of surface and groundwater, fresh and saline water, and other categories (USGS 2020).

Hydrologic studies can now access new data from satellites, which will improve the estimates of the global water balance. This enabled NASA (2015) to publish a study of annual fluxes of water by continent. The data show that heat from the sun evaporates 107 841 miles^3 (449 500 kilometers^3) of ocean water annually, about 20 times the contents of the Great Lakes. Precipitation on land is about 27 950 miles^3 (116 500 km^3) annually, with some 11 012 miles^3 (45 900 km^3) flowing through streams into the oceans and 16 938 miles^3 (70 600 km^3) evaporating to the atmosphere.

The water budget for the United States has also been studied extensively. Data show an annual average of precipitation of about 30 inches. The inflows and outflows include atmospheric fluxes and show outflows from surface and groundwater to the oceans, Gulf, Canada, and Mexico. Outflows to evaporation and consumptive uses are also shown.

In addition to the national level, water balances can be prepared for smaller political accounting units. For example, a water balance for a state or small nation might be needed for a water planning study. Cities and counties might use water balances to assess the need to change water policies. At a smaller scale, a district metered area in a city can be used to assess water uses and losses.

Water budgets for political jurisdictions are useful for planning purposes, but watershed and river basin boundaries do not coincide with political lines. This creates a continuing problem in solving water issues between jurisdictions. It is quite possible that if all state boundaries were on ridge lines, it might improve political cooperation.

Water Accounting in a Small Watershed and with a Reservoir

The water balance in a small watershed can be illustrated by a simplified stream segment with a reservoir. As shown in Figure 6.2, the reservoir stores water for release for municipal and industrial water in the city, for irrigation and for instream flows that serve water quality and environmental goals. Inputs will normally be stream inflow, precipitation, and any water imports to the reservoir. The outputs will be water releases, evaporation from the water surface, and seepage losses and any diversions.

A water balance for even this simple example requires a fair amount of data, mostly in time series for supply data to the reservoir at A, M&I demand data in the city, irrigation water demands as shown, and any loss data. The problems at the end of the chapter include a detailed example.

For any time increment such as a day, month or year, the change in reservoir storage will be given by:

$$I \times \Delta t - O \times \Delta t = \Delta S$$

where I denotes all inputs and O denotes all withdrawals and losses in the reservoir during the time increment Δt. The change in storage ΔS can also include changes in groundwater storage and soil moisture. The variables I and O are expressed as rates of change that become volumes when multiplied by the time increment. For example, if $I = 10$ cfs and $\Delta t = 1$ day, then $I * \Delta t = 10$ cfs-days or 19.85 acre-feet. If in the same day the outflow was, say, 10 acre-feet, then the net gain would be 19.85–10 = 9.85 feet. If the area of the reservoir water surface was constant at, say, 20 acres, then the change in water surface elevation during that day would be 9.85/20 = 0.49 feet.

Prior to the development of computers, mass curve approaches could be used to determine safe yield of a reservoir. The mass curve of demand is drawn on the same graph as the mass curve of supply, and the maximum gap between the two indicates the required storage capacity to meet the yield given by the slope of the

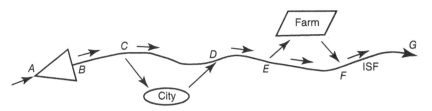

Figure 6.2 Water balance along a reservoir-stream system.

demand line. However, this method is too limited to use for reservoirs with time-varying demands and computer models work better.

A computer model provides a digital twin of a reservoir water balance. For a given time increment, say a month, the inflows, outflows, and changes in storage are simulated. The operating rules of the reservoir will affect change in storage, which will in turn determine rises and falls in the water surface. By setting the storage capacity and simulating a long period of record, whether the reservoir can supply a certain demand can be determined. Depending on the length of the period of record, the assumed starting volume of the storage contents may be important.

To provide an example of yield obtainable from reservoir storage, assume we can build a reservoir on a stream with a known safe yield on a weekly basis. Prior to the reservoir, the 7-day, 10-year low flow was some value, say 100 cfs. Now, we can smooth out the low flows by releasing stored water to increase them. In that way, the yield is increased to a value determined by analysis, say 150 cfs. To analyze this more precisely, we should use a storage-routing model and statistical input data.

Assume a reservoir with a capacity of 100 000 AF (123 MCM) is needed to increase safe yield from 100 to 150 cfs. The added safe yield of the reservoir is 50 cfs or 50 * 724 = 36 200 AF. This added yield has a probability that shows it is available except for once in 20 years, for example. This statistic will determine its value according to dependability.

If an AF is sold for $1000, the annual value added = $36.2 million. If building the reservoir costs $10 000/AF of storage, then total cost is $10 000 * 100 000 = $1 billion. Simple economics shows that if $i = 0.02$, then annualized cost = 0.02 * 10^9 = $20 million. Thus, $B/C = 36.2/20 = 1.8$. Of course, these numbers are dependent on I and the value of an AF of yield, and the example includes many assumptions.

For a complex system of streams and reservoirs, safe yield is the integrated product of the system. A system study and simulation are normally needed to determine safe yield of these systems. The rules of the system operation are important in determining yield because different sources can be selected at different times. As an example, in an irrigated area, the yields of combinations of water rights and ditch systems might be simulated to see how the yields of rights vary over a period of time.

Irrigation Water Balances

In irrigation, the "command area" is the area that can be reliably irrigated by a certain water source. This applies more to gravity systems because systems with pumps can move water around. Water balances for all irrigated areas involves

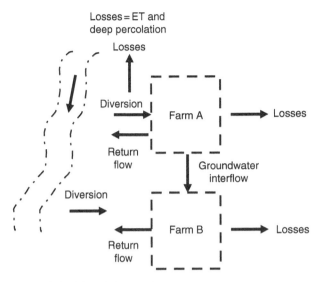

Figure 6.3 A farm irrigation layout.

variables that are important for efficiency but not easy to measure. This is particularly important for gravity systems that involve seepage and evaporation losses.

To illustrate a simple irrigation water balance, Figure 6.3 shows Farm A being irrigated by surface water from a stream. The delivery system will be subject to evaporation and seepage losses before the water is delivered to the farm. On the farm, there will be ET uses by crops and other plants and seepage losses. The seepage losses from all canals and surface areas may either remain as shallow groundwater and be available to downstream farms (such as B) or they may experience deep percolation as groundwater recharge to an aquifer. The diagram illustrates a common misconception about water waste. People may view a situation like Farm A's and think the water is wasted, but they do not know that Farm B depends on some of the "wasted" water that percolates from A as interflow.

The boundary for the water balance for a given area might be a single farm or an irrigation district with multiple farms. The relationships would be based on input minus output equals change in storage, and the storage could be soil moisture and local surface ponds. Each parameter will require measurement for accuracy and can only be estimated with large uncertainties. The exception might be the ET of the crops, for which there is a great deal of data. However, ET and evaporation depend on temperature and wind, which are highly variable.

Urban Water Balances

A water balance for an urban area can provide useful information on the origins and destinations of water as it moves through the area and changes in quality. The concept is based on an input–output model of diverse water fluxes, including wastewater. Such a study might be useful to develop the water footprint of the city. Also, the water-energy metabolism of a city might be studied to plan for improvements in heat flux and water demands.

In cities, the audit of a water distribution system to determine losses utilizes a water balance. These are needed to study losses and ways to improve system management. The approach can be illustrated by a water balance diagram that illustrates input volume divided into authorized consumption and losses (Figure 6.4). Losses are divided into categories of apparent and real losses, depending on whether they are mainly caused by errors and data problems or whether they are real leakage and other losses. An index is developed by dividing real losses by unavoidable losses.

Water Footprint and Embedded Water

The concepts of virtual or embedded water have been developed to inform decisions about water policy and management at different scales. A related concept is the water footprint, which is the total consumption of water of a person, business,

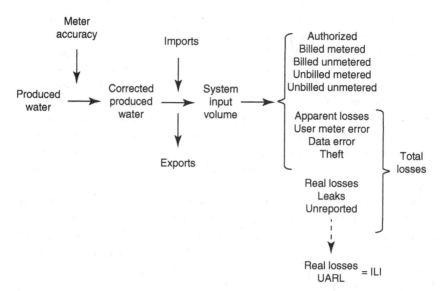

Figure 6.4 Water balance used to study losses in a distribution system.

or area, like a community or nation. Calculation of virtual or embedded water in a commodity, good, or service requires you to go to the steps of the production chain to determine how much water is embedded in products. Online calculators can provide general estimates.

Sometimes a discussion of water footprints will distinguish "green" water (water stored in soil and taken up by plants) and "blue" water (streamflow, lake water, and groundwater). At the site level, water can be separated into "white" (uncontaminated by human use), "black" (wastewater in contact with fecal matter or urine), and "gray" (wastewater not in contact with fecal matter or urine). Estimate of the quantities of the different colors of water will require measurements and calculations for any contextual situation.

References

NASA (2015). NASA balances water budget with new estimates of liquid assets. https://www.nasa.gov/feature/goddard/nasa-balances-water-budget-with-new-estimates-of-liquid-assets (accessed 15 May 2022).

Natural Resources Conservation Service (2022). Watersheds, hydrologic units, hydrologic unit codes, watershed approach, and rapid watershed assessments. https://www.nrcs.usda.gov/Internet/FSE_DOCUMENTS/ stelprdb1042207.pdf. (Accessed March 15, 2022)

University of Illinois (2020). The Earth's water balance. http://ww2010.atmos.uiuc. edu/(Gh)/guides/mtr/hyd/bdgt.rxml.

USGS (2020). Where is Earth's water? https://www.usgs.gov/special-topic/water-science-school/science/ where-earths-water?qt-science_center_objects=0#qt-science_center_objects

Questions

1 Define the water balance and give an example.

2 Write the basic equation of hydrology and define the terms. Give an example of how the equation is used in water balances.

3 Draw a simple water balance diagram and identify data required for computation of the water balance.

4 Name three uses of a water balance for decision-making.

5   Water rights determination, permitting of water uses, allocation of stream-flows, reservoir releases.

6   How does Horton's system of stream orders relate to watershed sizes?

7   What is the United States system to classify watersheds by size? What is a Hydrologic Unit Code?

8   Sketch a reservoir and show the relevant variables to explain how a water balance for it would be formulated.

9   Sketch an irrigated area and show the relevant variables to explain how a water balance for it would be formulated.

10   Sketch a water distribution system and explain how a water balance for it would be formulated. Explain how the use of water auditing relates to the water balance concept.

11   Identify and sketch a political jurisdiction as an accounting unit and explain how a water balance for it would be formulated.

12   What difficulty is created because boundaries of political jurisdictions and watersheds do not match?

13   Explain what is meant by blue water, green water, white water, black water, and gray water and how they relate to the water balance concept.

14   Explain what is meant by water footprint, embedded water, and virtual water how they relate to the water balance concept.

Problems

Transboundary Flow with Variable Flow and Annual Allocation

A river basin with a drainage area of $5000\,mi^2$ in a dry region produces an average annual flow of $10\,000$ acre-feet at the state line. An agreement was developed between the upstream and downstream states as follows: if the annual flow is between 25 and 75% of the average flow, water is divided 60% to the upper basin and 40% to the lower basin; if flow is greater than 75% of average, 100% of the excess over the average goes to the downstream state; if flow is less than 25% of average, the upstream state gets 100% of it. Compute the water allocations for the following flows. 20, 50, 75, and 150% of average.

The parties change the agreement to a basis of 10-year flow averages. If the allocation formula is 55% to the upper basin and 45% to the lower basin each year (based on the previous 10-year average), what will be the allocation for year 11 if the flows in AF for the previous 10 years are as shown in the table?

Year	Annual flow, AF
1	3141
2	9777
3	671
4	6285
5	5857
6	4653
7	3750
8	9843
9	2503
10	5485

Water Rights

A stream segment has four diversion points for water rights as shown (Figure 6.5). Under the appropriation doctrine of water rights (first in time, first in right), allocate the water to the four water rights for inflows at point A of 10, 20, and 30 cfs.

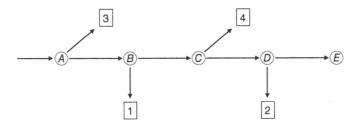

Figure 6.5 Sequence of water rights on a stream.

The water right seniority dates are:

Water rights (cfs)	Priority date	User
5.0	1880	Agriculture
12.0	1905	Industry
2.5	1908	Agriculture
4.2	1920	City

Losses in each stream reach, calculated as percentages of the flow at the head of the reaches are:

Reach	Losses (%)
A–B	5
B–C	12
C–D	9

Irrigation Losses

To determine the extent of seepage losses from an irrigation canal you will use a ponding method on a section of canal that is 1000 feet long. The cross-section is rectangular and 10 feet wide. Both ends of the section are sealed with temporary dams and the section is filled with water to a ponding depth of 6 feet. Evaporation rates are constant at 0.5 inches/day. A measurement of water level shows a decline of 2.0 feet in seven days. What is the loss rate in gallons per square foot per day?

Urban Water Audit

Using the information in the chapter about auditing water distribution systems, compute the Infrastructure Leakage Index for a system with the following data:
Corrected input volume = 6000 MG
Total authorized consumption = 5400 MG
Total losses = 1393 MG
Total apparent losses = 357 MG
Total real losses = 1035 MG
Unavoidable annual real losses = 242 MG

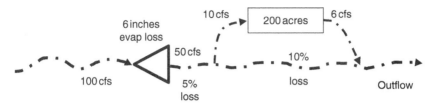

Figure 6.6 Stream-reservoir system.

Reservoir Water Balance

The stream system shown has the flows and losses given for a 30-day month. The 100 cfs inflow is into the reservoir; the 50 cfs is the reservoir release; the 10 cfs diversion is from the stream to the farm without any conveyance losses; and the 6 cfs outflow is from the farm to the stream without conveyance losses. The reservoir has a surface area of 500 acres. The losses shown in the stream are applied to the flows at the beginning of the reaches. Compute for the given month: (i) change in water surface elevation in the reservoir; (ii) sum of usage and loss on the 200 acre farm field in acre-feet and feet of water; and (iii) average outflow at the end of the reach in cfs. If you lack any data, state your assumptions, and make the computations (Figure 6.6).

Reservoir Water Balance

A reservoir with maximum capacity of 100 000 acre-feet has a surface area of 5000 acres, which is constant and does not change with water elevation. The volume in the reservoir on 1 January of a given year is 28 885 acre-feet. During the year: the lake evaporation is 70 inches, the seepage loss from the reservoir is a constant 10 cfs, the inflow averages 200 cfs, and the outflow averages 100 cfs. What will be the volume at the end of the year on 31 December?

Small Basin Water Balance

A small watershed with an area of $10\,\text{mile}^2$ has 40 inches of precipitation in a year. There is an interbasin transfer during the year, some 5000 acre-feet of water are exported out of the basin. The gage at the discharge point from the watershed shows an annual average of 18 cfs. Create a water balance for the watershed during the year, compute the quantities, and explain any parts of the water balance that might not be given in the data here.

7

Flood Studies: Hydrology, Hydraulics, and Damages

Flood Hazards and Water Management

As a serious threat in cities and other vulnerable areas facing extreme events, flooding is often cited as the world's most costly natural disaster. To reduce and mitigate flood risk, water managers require knowledge of hydrology, hydraulics, and how flooding causes damage and injuries. This chapter explains these topics along with a brief technical discussion of flood mapping, which depends on the hydrologic and hydraulic analysis and relates flood extent to risk. The focus of this chapter is on technical aspects, and it is complemented by Chapter 21, which explains policies and tools for flood risk management.

Hydrologic Causes of Floods

The main driving force causing flooding is extreme precipitation, including rainfall and snowmelt, which can also cause flooding and exacerbate rainfall-induced floods. Rainfall on melting snow can increase flooding due to combined snowmelt and high rates of runoff from the rain. Extreme precipitation falling on bare soil or impervious areas will generate runoff more quickly and at higher rates than thickly vegetated areas. Antecedent conditions may leave soils saturated and reduce infiltration rates. In areas with steep slopes, flooding occurs more rapidly than in flat areas. In urban areas, impervious surfaces exacerbate flooding.

These diverse conditions lead to different types of floods. Steeper areas subject to intense convective storms have flash flooding. In flat areas like the Mississippi River Valley, big river floods rise slower and last longer. In urban areas, flood magnitudes vary from minor to major levels. In coastal areas flooding occurs due to storm surge and heavy precipitation.

Water Resources Management: Principles, Methods, and Tools, First Edition. Neil Grigg.
© 2023 John Wiley & Sons, Inc. Published 2023 by John Wiley & Sons, Inc.

Extreme precipitation that causes flooding can occur from different types of storms. The National Oceanic and Atmospheric Administration provides abundant information about storm types, and descriptions are provided by the NOAA National Severe Storms Laboratory (https://www.nssl.noaa.gov/education/svrwx101/floods/forecasting).

The main mechanisms that cause precipitation are cooling and condensation. Convective storms occur where surface heating causes upward motion that transports available moisture to cooler layers of air. The lifting of moist air can also occur with transport over mountain ranges, which causes orographic precipitation. When moisture moves around due to differential pressures, it causes cyclonic storms. When electrical activity is present, convective motion can create thunderstorms. Storms may occur with different numbers and configurations of rainfall cells. They can be single- or multi-cell storms, where individual cells last short periods and the system lasts longer.

Flood Frequency Analysis

The threats from flooding are analyzed statistically in flood frequency analysis, where the goal is to estimate the likelihood of peak flow magnitude and other attributes of the runoff hydrograph. The technical literature on flood frequency analysis is extensive, and a good introduction to the techniques is from Leo Beard (1962), who led in developing many hydrologic methods for the USACE. The Beard publication is considered a classic and worth adding to a professional library.

Flood frequency analysis is used for peak flow estimation in analysis of flooding in channels and for dam spillway assessment. In some cases, a return period is needed whereas in others an extreme value like the probable maximum flood (PMF) due to probable maximum precipitation (PMP) is needed. The USACE has also used a Standard Project Flood (SPF) in the past, which represents the extreme flood that a project is designed for, considering the most severe combination of meteorological and hydrologic conditions that may be expected.

The most common problem is to determine the statistics of extreme flooding at a point, which is becoming more complex due to the possibility of climate change and resulting non-stationarity. The historical record is used as a guide if it is available, and historical accounts by residents and publications can also be useful. The basic procedure is to conduct a statistical analysis of the floods of record. The flood frequency analysis uses extreme value theory for precipitation and stream flow. Precipitation data were explained in Chapter 5 to illustrate rainfall intensity–duration–frequency (IDF) curves and rainfall magnitudes on an hourly or other basis.

Flood researchers have developed methods to apply statistical distributions to flood runoff. These are like low flow frequency analysis, which was discussed in

Chapter 5. The basic procedure is to take a series of annual peak floods and order the values from lowest to highest ($m = 1-n$). You then assign plotting positions ($P = m/n + 1$). You can plot these on probability paper and select values, even with limited extrapolation, but the preferred approach is to fit a probability distribution, like the log-Pearson Type III that is recommended in Bulletin 17C, which is now published by USGS (England et al. 2019).

Current recommendations go beyond those in the initial versions of the guidance. They include considering all data available, including historical, paleoflood, and botanical data, as well as evaluating various ways to interpret and extend statistical parameters.

A common flood risk problem is to compute the probability (J) that within n years a flood of return period T (with probability of exceedance P in a year) will occur. If this is approached directly, it is infeasible to solve it because you would start with the probability that it occurred in year 1, then consider years 1 and 2 together, and so on up to n years. A convenient way to solve the problem is to consider the probability that the flood does not occur in a year, or $1 - P$. If, for example, you consider 2 years for a 100-year flood, this would yield $(1 - 0.01)^2$ as the probability that the flood would not occur in 2 years. The probability that it would occur is then, $J = 1 - (1 - 0.01)^2 = 0.0199$. This computation can be extended by replacing 2 in the equation with the number of years, for example, for $n = 50$ years, $J = 1 - (1 - 0.01)^{50} = 0.395$.

Computation of Storm and Flood Runoff

Storm runoff occurs due to excess precipitation as modified primarily by interception, infiltration, and depression storage (see Chapter 5). Small watersheds contribute flow to stream channels, which collect runoff with flood magnitude rising as the watershed area increases going downstream.

While there are many sophisticated theories about the rainfall-runoff process, the basic mechanism is explained by the Rational Method, which is relatively simple compared to advanced models. Although the method is simple, a few complexities must be considered to use it effectively. The equation is:

$$Q = CiA,$$

where C is a dimensionless runoff coefficient, i is the rainfall intensity in inches per hour for a selected return period that corresponds to the time of concentration of the watershed, and A is the basin area in acres. This convenient formula works adequately for small watersheds, up to about 200 acres. In the English system of units this yields Q in cfs because the units in inch-acres per hour turn out to be nearly equal to cfs.

For larger watersheds, the Rational Method does not work, and different methods are needed, either based on models or a unit hydrograph approach. In small watersheds, even up to a few square miles in area, land forms and vegetation can be relatively uniform and runoff phenomena like that for a very small basin can occur. In larger basins, it is common that changing landforms, vegetation, storage receptacles, and other influences will modify runoff and require more complex modeling.

Hydrographs

The runoff hydrograph is the most common tool for analysis of the time distribution of flood magnitudes. It is a graph of flow versus time at a point and is applicable to basins of all sizes. Many methods for use of hydrographs have been developed.

A simplified view of a hydrograph is shown in Figure 7.1. The basic parameters are flow, time, time-to-peak, recession time, and base flow. In particular, base flow can be more complex than shown due to infiltration and slow movement of groundwater. Base flow is shown as linear and increasing slightly, but there is no way to measure it exactly and it is not always well-defined.

Unit hydrograph theory was developed to provide a mechanism to analyze storms with different characteristics to create runoff hydrographs. The word unit means that the volume under the hydrograph is equivalent to 1 inch of water over

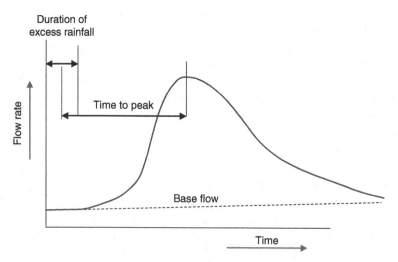

Figure 7.1 Flood hydrograph parameters.

the basin. For example, if the basin area is 1000 acres (40 ha), the runoff volume expressed by a unit hydrograph would be 1000 acre-inches or 83.3 acre-feet (102 000 CM). Deriving a unit hydrograph for a basin requires analysis of historical storms, which provide information about watershed responses. Generally, a storm that is relatively uniform over the basin with the duration of about one third of the basin lag time provides a characteristic storm to derive the unit hydrograph. The total runoff would be converted to 1 inch equivalent to create unit hydrograph.

Once the unit hydrograph is derived, you can utilize rainfall pulses to create successive hydrographs from them to obtain an integrated hydrograph for an entire storm. This enables the computation of runoff from complex storm patterns. Hydrographs can also be combined to sum contributions from different parts of a large watershed. Methods to apply unit hydrographs are explained in references like the National Engineering Handbook published by the Natural Resources Conservation Service (2022) and is available for downloading.

A simple triangular hydrograph (Figure 7.2) can be used with the Rational Method to study approximate runoff volumes and the operation of detention basins. For a triangular hydrograph, the peak is the runoff computed by the Rational Method formula and the base is determined by estimates of rainfall rates and durations. For example, the time-to-peak could be set equal to the time of concentration or assuming that the rainfall started initially, then the flow rose linearly in intensity to the peak value, after which it started to decline. Similar assumptions can be made for the recession part of the hydrograph. The figure shows the concept, with the recession limb at twice the time of concentration. These assumptions must, of course, be tested and the validity of results must be discussed in any reports.

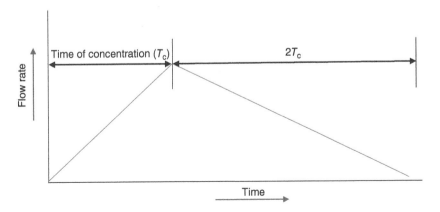

Figure 7.2 Triangular hydrograph.

Flood Hydraulics and Conveyance

After flow magnitudes and durations are known, the next step is analysis of the conveyance capacity of the channel or pipeline system. The methods of flood hydraulics provide the main tools for analysis of flow in channels to convey runoff. These methods are explained in detail in references such as the NRCS National Engineering Handbook, mentioned earlier.

Major flows will be conveyed in channels and smaller flows may be conveyed in stormwater pipes. In the case of channels, flood hydraulics considers floodways as comprising main channels and floodplains. Figure 7.3 explains how the National Flood Insurance Program defines a main channel and floodplains. The designated flood elevation, usually the 100-year, is allowed to rise one-foot after obstructions of the floodway fringe area are completed. Of course, floodplains include many land uses, including natural areas.

If a flood is long-lasting, flow in the floodway may be near uniform in depth and velocity down the valley, but the early and later stages of flow will be nonuniform and perhaps unsteady as shown by advancing and retreating flood waves.

Uniform flow is often useful as a first approximation of channel flow and can be analyzed with the Manning equation, which is:

$$Q = (1.49 / n) * A * R^{2/3} * S^{1/2},$$

where Q is the discharge in cfs; n is a roughness factor; A is the cross-sectional area; and S is the channel slope downstream.

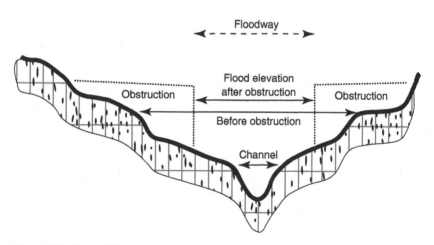

Figure 7.3   Channel floodway.

Nonuniform flow requires a different computational approach, which is facilitated by computer models. Unsteady flow to analyze a flood wave always requires computer modeling that includes solution of differential equations. Models to analyze the flows are explained briefly in Chapter 9.

The analysis of hydraulics in stormwater pipe systems requires consideration of free surface and pressurized flow. Small flows will normally have free surfaces and can be analyzed in a similar way as channel flow, although analysis usually involves approximations due to changing conditions and flows inside the pipes. Higher flows that surcharge the pipes require a different approach to consider pressure effects. As in many open channel cases, use of computer models facilitates analysis.

Bridges and other channel obstructions create backwaters that exacerbate flooding. Analysis of flood risk must consider how bridges affect flows. The Federal Highway Administration has developed a guide for considering them, along with many other computational aids.

Reservoir Storage-Routing

Reservoirs are important for control of flooding and must be operated effectively to take advantage of storage to mitigate flood damages. Risk of dam failure during flooding must be considered, which means that the hydraulic capacity of spillways must be adequate. The USACE has many responsibilities for operating reservoirs during flooding, and provides guidance through its Engineering Manuals. USACE (2022) publication EM 1110-2-3600 entitled "Management of Water Control Systems" explains policies for reservoir operations.

Routing of a flood through a reservoir is illustrated by Figure 7.4. As the discharge increases, the reservoir begins to capture flood water such that its discharge is less than the flow into it. The release from the reservoir is regulated through operational controls. The difference between the inflow and outflow is storage of water into the reservoir. After the hydrograph discharge falls below the reservoir release curve the reservoir will be emptying.

Models are useful in studying reservoir operation strategies. They are discussed in Chapter 9 on a general basis, and a simple example of reservoir routing is presented here. To simplify, we use a detention pond in an urban setting. Figure 7.5 shows the pond with inflow and outflow hydrographs. The routing comprises converting the inflows to outflows through consideration of storage in the pond.

In the example, the inflow hydrograph is approximated by a simple triangular form like the one shown in Figure 7.2. Peak flow is computed as 20 cfs and the time of concentration is 20 minutes, with the recession limb of the hydrograph reaching zero at 60 minutes.

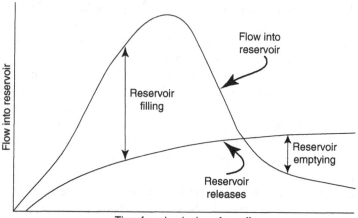

Flow into reservoir

Reservoir filling

Reservoir releases

Reservoir emptying

Flow into reservoir

Time from beginning of runoff

How a reservoir captures flood water

Figure 7.4 Hydrographs of flood routing through a reservoir.

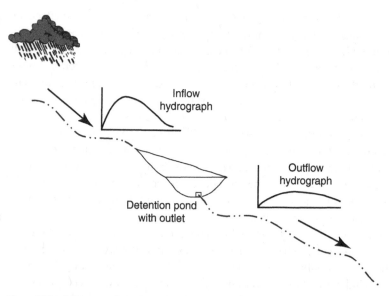

Inflow hydrograph

Detention pond with outlet

Outflow hydrograph

Figure 7.5 Inflows and outflows from a stormwater detention pond.

The pond has a surface area of 5000 feet^2, which we assume as constant for simplicity. In a more complex example, we could consider the variation of area with water elevation. Outflow from the pond is controlled by a 2-foot wide opening that operates like a sharp-crested weir with the equation:

$$Q = 3.33LH3/2$$

where Q is the discharge, L is the width (2 feet for this example), and H is the depth in the pond.

Routing is shown in this table. The variables are defined below the table.

t^a	Q_{In}	V_{In}	Q_{out}	V_{out}	DS	S	H
0	0	0	0	0		0	0
2	2	120	0	0	120	120	0.02
4	4	360	0	1	359	479	0.10
6	6	600	0	7	593	1073	0.21
8	8	840	0	26	814	1887	0.38
10	10	1080	1	66	1014	2901	0.58
12	12	1320	1	135	1185	4086	0.82
14	14	1560	2	236	1324	5410	1.08
16	16	1800	4	372	1428	6838	1.37
18	18	2040	5	544	1496	8333	1.67
20	20	2280	7	749	1531	9864	1.97
22	19	2340	9	984	1356	11220	2.24
24	18	2220	11	1225	995	12215	2.44
26	17	2100	13	1435	665	12880	2.58
28	16	1980	14	1589	391	13271	2.65
30	15	1860	14	1690	170	13441	2.69
32	14	1740	15	1745	−5	13437	2.69
34	13	1620	15	1761	−141	13296	2.66
36	12	1500	14	1747	−247	13049	2.61
38	11	1380	14	1709	−329	12720	2.54
40	10	1260	14	1653	−393	12327	2.47
42	9	1140	13	1584	−444	11883	2.38
44	8	1020	12	1506	−486	11398	2.28
46	7	900	11	1420	−520	10878	2.18
48	6	780	11	1329	−549	10329	2.07

(Continued)

t^a	Q_{in}	V_{in}	Q_{out}	V_{out}	DS	S	H
50	5	660	10	1234	−574	9755	1.95
52	4	540	9	1138	−598	9157	1.83
54	3	420	8	1040	−620	8537	1.71
56	2	300	7	941	−641	7896	1.58

[a] t = time in minutes; Q_{in} and Q_{out} = discharges into and out of the pond in cfs; V_{in} and V_{out} = volumes of water into and out of the pond during the time increments in cubic feet; DS = change in storage in cubic feet; S = ending storage in cubic feet; and H = depth of water in the pond in feet.

The table illustrates two-minute time increments, which produce a result as shown in Figure 7.6. Using a shorter time increment will increase the accuracy of the simulation. Decreasing the size and capacity of the outlet structure would show a more dramatic decrease of the outlet flow, but would increase the water elevation in the pond.

Flood Forecasting

Forecasting flood magnitude depth and timing is needed to alert residents about hazards and to plan reservoir releases. Forecasting is more difficult for flash floods than for slow rising floods because of the shorter response times. The NOAA National Weather Service has a flood forecasting service that utilizes advanced

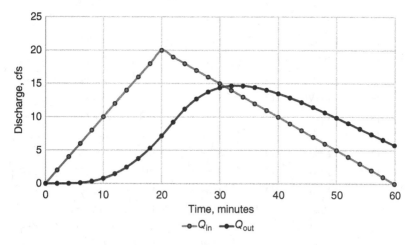

Figure 7.6   Inflow and outflow of detention pond.

techniques such as radar and computer models to make forecasts. It works through its Weather Forecast Offices and River Forecast Centers in partnership with federal, state, and local agencies to utilize computer models and data from various sources. The hydrologists who perform flood forecasting draw from an array of data comprising river stages, rainfall, and local reports to formulate their forecasts. Over the years, river models have improved, and forecasts have become more reliable. In 2016, NOAA introduced a National Water Model that is intended to advance the state of the art of forecasting with much better models. It is described in Chapter 9.

Flood Mapping

Flood mapping is a way to outline risk zones to enable property owners, insurance companies, and local governments to balance responsibilities for flood protection. Preparation of flood maps requires analysis of topography, land use, hydrology, and hydraulics. Flood maps evolved from the beginning of the flood insurance program, and their complexity has increased since that early beginning. Because flood maps carry implication of rate levels and because of uncertainties they are controversial politically.

To prepare a flood map, engineers work with local governments to outline areas and scopes of work. Hydrologic analysis prepares the flows to be considered and hydraulic analysis develops depths of flow for floods in different locations of different magnitudes. The mapping involves locating how the water will spread, which requires analysis of complex land uses which may be uncertain and may change. Riverine and coastal flood zones are shown differently. Definitions of the risk zones are available in the Flood Insurance Manual from FEMA (2022). This downloadable manual contains extensive information about flooding and nonstructural solutions, and it is updated periodically by FEMA.

Flood Damages

Flood damages are caused primarily by inundation and are affected by velocity as well. Their potential must be assessed in the planning process and, while data on flood damages involves many parameters, guidelines have been improving. The FEMA software tool HAZUS has some data embedded in it, but definite studies need more local information. The HAZUS Technical Manual has extensive discussions of flood damage data that has been collected to support actuarial studies. These were compiled by FEMA's Flood Insurance Administration, which is now called the Flood Insurance and Mitigation Administration.

Flood damages as a function of inundation differ among building types. The simplest depiction is for a one-story residential home with a basement where a

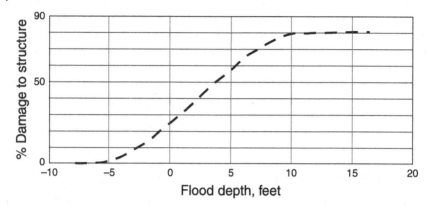

Figure 7.7 Typical depth–damage curve for single family residential structure.

typical curve would take this general shape shown in Figure 7.7. Flood depth is measured from the first floor elevation. Note that damage begins to occur before the water level reaches the first floor. This would be the case even if there was no basement because of soil upheaval and damage to foundations.

Dam Breaks

Historically, there have been many dam failures with severe and sometime catastrophic consequences. These are reviewed in Chapter 21, which explains how the 1889 failure of a dam in Johnstown, Pennsylvania caused many casualties and set the stage for legal reform and in 1976, failure of the USBR's Teton Dam was a wake-up call for the national dam safety program. Many other dam failures have occurred in the United States and globally, and these continue.

Protecting dams from risk of failure requires hydrologic, hydraulic, geotechnical, and structural analysis. The Association of State Dam Safety Officials publishes extensive information on dam safety and how to conduct the analyses. Further information is available in Chapter 3 about dam infrastructure and in Chapter 21 about dam safety.

Environmental Benefits of Floods

Flooding is a natural phenomenon and has positive environmental effects on riparian zones, floodplains, and wetlands. It inundates wetlands with nutrients and contributes nutrients to lakes and streams to support fisheries.

River sediments provide nutrients to topsoil and fertilize agricultural lands. Soil deposition builds land and creates natural levees. Floodwaters also recharge groundwater.

References

Beard, L.R. (1962). Statistical methods in hydrology. https://www.hec.usace.army.mil/publications/TrainingDocuments/TD-4.pdf (accessed 15 May 2022).

England, J.F. Jr., Cohn, T.A., Faber, B.A. et al. (2019). Guidelines for determining flood flow frequency – bulletin 17C (ver. 1.1 may 2019). In: *U.S. Geological Survey Techniques and Methods*, book 4, chap. B5, 148 p. Geological Survey U.S. https://doi.org/10.3133/tm4B5.

FEMA (2022). Flood insurance manual. https://www.fema.gov/sites/default/files/documents/fema_nfip-all-flood-insurance-manual-apr-2021.pdf (accessed 15 May 2022).

Natural Resources Conservation Service (2022). *National Engineering Handbook*. National Resources Conversation Service https://www.nrcs.usda.gov/wps/portal/nrcs/detail/?cid=nrcs141p2_024573.

USACE (2022). Management of water control systems. https://www.publications.usace.army.mil/Portals/76/Publications/EngineerManuals/EM_1110-2-3600.pdf (accessed 15 May 2022).

Questions

1 What are the main differences between river floods and flash floods?

2 What units of precipitation data are used in analysis of stormwater or flood control runoff?

3 Explain why flooding due to rainfall on melting snow tends to be greater than on bare soil.

4 Define flood "return period" and explain how it relates to "exceedance probability."

5 What is the difference between a probability distribution and an extreme value probability distribution?

6 What probability distribution is recommended for flood frequency analysis? Why?

7 Explain: Standard Project Flood, Probable Maximum Flood, and Probable Maximum Precipitation.

8 Explain how historical, paleoflood, and botanical data can improve estimates of flood frequency.

9 Write the equation for the Rational Method and define the terms.

10 Draw a typical hydrograph and label the axes. Identify the rising limb, peak, and recession limb.

11 Draw a unit hydrograph and show the rainfall pulse that goes with it. Explain how rainfall duration is considered when using a unit hydrograph.

12 How does a flood plain change the hydraulics of a river flood when overbank flow occurs?

13 Write the Manning Equation and define the terms.

14 Define unsteady flow and explain how its theory is used to measure movement of a flood wave.

15 Draw a valley profile and show how a dam break will affect flows and depths downstream from the failed structure.

16 How are bridges vulnerable to water flow and hydraulic forces, and what are their effects on flood backwater?

17 Explain how a reservoir can reduce downstream flood magnitudes.

18 What is the main cause of flood damages? Where can data about them be found for planning purposes?

19 Levees are key features of flood protection systems in low-lying areas. How do they affect flood hydraulics? What happens when a levee breaks?

20 Sketch a plan view of a flood insurance map and identify its features.

21 What are the main environmental benefits of flooding?

Problems

Simple Runoff Volume and Units

In flooding, the streamflow fluctuates rapidly due to storm runoff. If you have a watershed of 10 mile² in tributary area and effective storm runoff is 1 inch, what will be the flood runoff in acre-feet? If this runoff entered a reservoir with a 250-acre surface area, how much would the water level rise?

Detention Pond

A small detention pond experiences a flood inflow that increases linearly from 0 to 1000 cfs in two hours. Then it decreases linearly to zero again in six more hours. How much water flows into the pond?

Storage Required in a Flood Reservoir

A stream is flowing at 10 000 cfs prior to a heavy rain. A flood wave causes streamflow to increase from 10 000 to 20 000 cfs in 24 hours. Then it peaks and drops back to 10 000 cfs in the next five days. Estimate the volume of flood water that passes.

Rational Method

The Rational Method is effective for estimating runoff from small watersheds up to about 200 acres. Assume two converging small watersheds (Figure 7.8) with runoff coefficients of 0.5 as shown. Watershed A has an area of 75 acres and a time of concentration of 15 minutes and watershed B has an area of 200 acres and a time of concentration of 30 minutes. The IDF data show a rainfall intensity of

Figure 7.8 Converging watersheds.

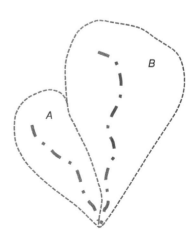

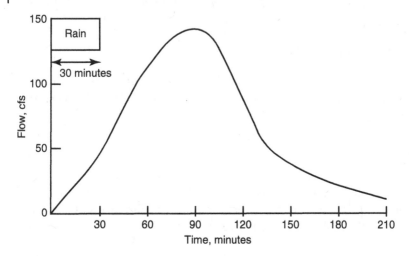

Figure 7.9   Flood hydrograph for conversion to unit hydrograph.

2 inches/hour for a 15-minute duration and 1.6 inches/hour for 30 minutes. Compute the peak flows of each watershed. Use these peak flows to sketch triangular hydrographs with the rising component equal to the time of concentration and the recession component at twice that time. What will be the peak flow of the combined runoff of the two watersheds?

Hydrograph

Flood runoff from a 1000-acre watershed and resulting from a 30-minutes storm with 1-inch of precipitation is shown by the graph (Figure 7.9). Convert the hydrograph to a unit hydrograph and label the ordinates.

Using the unit hydrograph you developed, compute the discharges from the watershed for a 2-hour storm with rainfall amounts in 30-minute increments equal to 0.5, 0.75, 1.1, and 0.4 inches.

Return Period and Exceedance Probability

Flooding is described in terms of probability and return period. If a storm that produced a 1 inch of runoff is described as a 10-year storm, what will be the probability in any given year that it will be equaled or exceeded?

Risk of Flooding

What is the probability that a 50-year river level will occur within the next 10 years at any given location?

8

Water Quality, Public Health, and Environmental Integrity

Introduction

Water managers have responsibilities to protect water quality, public health, and environmental integrity, which are linked in frameworks such as Environmental Health, One Health, and similar holistic concepts. Whereas in the past, it might have been sufficient to approach these responsibilities separately, it has become clear that preventing failures due to poor decisions requires integrated approaches.

Water managers should understand how their roles in water resources decision making affect issues of health and the environment. They should know how water quality is measured, how it changes in the hydrologic cycle, how point and nonpoint pollution occur, and how water quality affects people and ecosystems.

Knowledge domains that support these responsibilities are water quality management, environmental science and engineering, and public health engineering. These evolved separately but have many connections. Regulatory programs for them also tend to be separate, but decisions should integrate their goals to the extent possible. This goal is embodied in the rising interest in integrative management, see Chapter 23. Legislation and regulatory controls may be along separate tracks, so it is up to water resource managers to find ways to facilitate integration even among stove-piped regulatory systems.

This chapter begins with a short explanation of the water–health–environment nexus, or the space where these three management areas intersect. Then it provides a definition and explanation of environmental and drinking water quality and reasons for its degradation. The discussion provides background for Chapters 15 and 20 that explain the legal and procedural aspects of water quality management. The linkages between water management, exposure pathways, and disease incidence are then discussed in the context of public health, and this is followed by an explanation of environmental integrity as a metric

Water Resources Management: Principles, Methods, and Tools, First Edition. Neil Grigg.
© 2023 John Wiley & Sons, Inc. Published 2023 by John Wiley & Sons, Inc.

that encompasses water quality and environmental sustainability. Methods to assess stream health are discussed in relation to the concept of ecohydrology (see Chapter 5).

Water–Health–Environment Nexus

How water quality, public health, and environmental integrity are linked can be seen through the lenses of systems thinking, integrated management, and the nexus concept. The conceptual frameworks of these management tools are discussed in Chapter 23. The linkages are evident in the Sustainable Development Goals (SDGs) of the UN and through the integrative framework of One Health, which organizes the interconnections between people, animals, plants, and their shared environment.

Figure 8.1 illustrates the 17 SDGs (bottom of diagram) and the three pillars of One Health, environmental health, human health, and animal health (top of diagram). The SDGs are listed in Table 8.1.

Not all objectives of the SDG clusters are captured by the single words, and a list of them is provided below. The significance of the display is to show how one or all of the three pillars of One Health are integrally related to each SDG. Water as a resource also links to each of the three One Health pillars, its relationships to the 17 SDGs are also evident. Whereas only one of the SDGs is labeled with water, it is obvious that all link to it, like energy dependence on water or the water footprints of cities, for example.

Drivers of Pollution and Threats to Public Health and Ecosystems

Water pollution threatens health and life and is especially hard on the poor and vulnerable. Environmental degradation also causes loss of biodiversity and ecological services. Drivers of change that increase urgency of these threats are global development, climate change, and rising aspirations that create pressures on water systems and services. These can be illustrated in a DPSIR framework like the one illustrated in Chapter 1.

Human-caused change and natural forces cause pollution of environmental water. The human causes focus on point and non-point sources of activities such as urbanization and wastewater discharge, industrial discharges and land uses, farming, mining, energy use, stream-channel alteration, and animal-feeding cause pollution. Nitrogen and phosphorus dissolve in runoff creating excess nutrients that promote algae growth, low oxygen, and fish kills. Chemicals such as

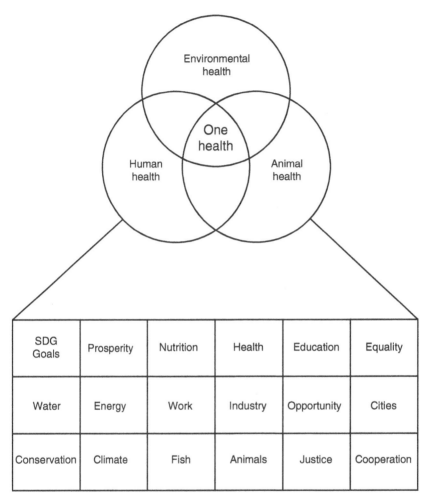

Figure 8.1 SDG goals linked to One Health framework.

pharmaceutical drugs, dry-cleaning solvents, gasoline, and pesticides find their ways into water bodies through various paths.

Water quality changes due to natural forces vary by location, seasons, climate, and geology. Runoff dissolves and transports minerals and salts such as calcium and chlorides, plant nutrients such as nitrogen and phosphorus, and trace elements such as selenium, chromium, and arsenic. It percolates through organic roots and leaves, and it reacts with microscopic organisms. Water may carry plant debris and sand, silt, and clay to rivers and streams making the water appear muddy. When water evaporates from lakes and streams, the concentration of

Table 8.1 SDG goals.

SDG		SDG goal
1	Prosperity	No poverty
2	Nutrition	Zero hunger
3	Health	Good health and well-being
4	Education	Quality education
5	Equality	Gender equality
6	Water	Clean water and sanitation
7	Energy	Affordable and clean energy
8	Work	Decent work and economic growth
9	Industry	Industry, innovation and infrastructure
10	Opportunity	Reduced inequality
11	Cities	Sustainable cities and communities
12	Conservation	Responsible consumption and production
13	Climate	Climate action
14	Fish	Life below water
15	Animals	Life on land
16	Justice	Peace and justice strong institutions
17	Cooperation	Partnerships to achieve the goals

dissolved minerals increases. Landslides, floods, volcanic eruptions promote run-off of sediments, volcanic ash, and salt discharges.

Impairment of drinking water quality is caused by threats to source water, failed treatment processes, and distribution systems that allow pollutants to contaminate the water. In many places, self-supply systems become polluted through seepage of bacteria or chemicals into groundwater. Other places lack access to safe drinking water in the first place.

Measuring Water Quality

While it is easy to spot water pollution, it can be difficult to measure water quality because there is no universal definition or index, and it involves multiple criteria. The constituents that determine water quality are linked directly to human and environmental health. While people, plants, and animals require certain levels of them, the problems occur when the levels of the constituents get out of balance.

Descriptors such as "good," "bad," or "impaired" water quality require a standard to be judged. In that sense, the condition of water is measured relative to its intended uses for human or environmental purposes. This condition requires numerous physical, chemical, and biological metrics to describe. The Clean Water Act (see Chapter 20) embodies that philosophy. How water quality standards are set differs among environmental water and drinking water. Chapter 20 also explains how these standards are set to protect environmental water for intended uses and to manage drinking water to protect public health.

Physical, chemical, and biological constituents in water require parameters to measure water quality. Physical parameters include sediment concentration, temperature, taste, odor, and color, among others. Chemical parameters include oxygen content, inorganic chemicals, organic chemicals, and radionuclides. Oxygen content measures the capacity of water to sustain aquatic life. Inorganic chemicals include salts, nutrients, and metals such as lead, mercury, and selenium, among others. Nutrients are important indicators of stream and lake health, especially carbon, nitrogen, and phosphorous. Organic chemicals in water include pesticides and industrial chemicals. Radionuclides can include isotopes from nuclear activity in power plants, research, and medical activity. The biology of water involves healthy organisms such as plankton and unhealthy organisms such as bacteria, algal toxins, and viruses.

Public Health Impacts of Water Contamination

The health impacts of unbalanced water quality have always been evident as people got sick and died from drinking contaminated water. Today, we recognize broader dimensions of how water relates to public health, which is a holistic concept that means the health or well-being of the whole population. It is distinguished from the branch of medicine which refers to treatment. The main topics of public health include hygiene, epidemiology, and disease prevention. Epidemiology deals with the incidence, distribution, and possible control of diseases and other factors relating to health. Its methods focus on surveillance and studies of determinants of disease. Some work of epidemiologists and environmental health specialists overlaps with that of environmental engineers.

Prior to the emergence of modern water quality management, waterborne diseases such as typhoid and cholera devastated communities. Responses to drinking water threats were provided through engineering and public health professions. A combination of these responses might be called hydro-epidemiology, or water-related epidemiology. Taken together, these responses fit into the descriptor of "public health engineering," which is explained in Chapter 2.

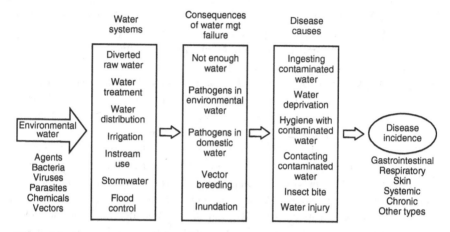

Figure 8.2 Water-related pathways to contamination and disease.

While much progress has been made, public health impacts of degraded water quality still occur, ranging from diarrheal to chronic diseases caused by failed drinking water systems, and including various skin and systemic diseases caused by contaminated environmental and irrigation water. These are particularly difficult in low-income settings where people lack access to protective systems.

An integrated model can show the effects of water management on health. A starting point to illustrate is the "F-diagram," which is used by public health specialists to illustrate pathways from pollution to health impacts. The diagram's name comes from elements that start with the letter F: feces, fingers, flies, fields, fluids, food, and faces. It considers all exposure media, such as air, water, food, work areas, through which contamination can occur.

Figure 8.2 shows a version of an F-diagram to illustrate how vectors of disease move through water-related pathways to expose people to different kinds of diseases.

For the public health community, a classification of these has been developed with four categories of pathways (water-borne, water-washed, water-based, and water-related) (Eisenberg et al. 2001). Additional categories can be added to describe water scarcity (water-deprived category) and to capture flood injuries (water-damaged category). These are combined in Table 8.2 to illustrate the pathways, water management failure types, exposure mechanisms, and types of diseases and consequences.

Global Problem of Safe and Reliable Water

The global water and sanitation issue is high on lists of priorities among the international community and has been assigned as one of the SDGs. Of the world's some eight billion people, the majority lack access to safe, reliable, convenient, and

Table 8.2 Water pathways, exposure mechanisms, and diseases.

Pathways	Management failure	Exposure mechanisms and diseases
Water-borne	Safety and access to drinking water	Ingestion of contaminated water with bacteria or viruses. Dysentery and diarrheal diseases
Water-washed	Limited access to sanitation	Poor hygiene. Scabies, trachoma and flea-, lice-, and tick-borne diseases, and waterborne diseases
Water-based	Stream water quality degradation	Parasites in water organisms. Dracunculiasis, schistosomiasis, and other helminths
Water-related	Stormwater and standing water	Insect vectors. Dengue, filariasis, malaria, and yellow fever
Water-deprived	Water availability and scarcity	Lack of or marginal supplies. Poor sanitation. Dehydration, hunger, increased vulnerability
Water-damaged	Failed flood management systems	Floodwater impacts. Injuries, death, cholera, weakened disease resistance

affordable drinking water. Billions are also exposed to pathogens that follow other pathways, including contact with contaminated water, personal hygiene, and vectors breeding in water. Their exposure to water scarcity and flood inundation also poses threats to public health. The problems are driven more by economics, governance, and social conditions than by technical difficulties. Water managers know how to provide safe water, but overcoming the many obstacles is difficult.

The extent of the global water and sanitation issue is reported by the Joint Monitoring Program for Water Supply and Sanitation (JMP), which is sponsored by the World Health Organization (WHO) and UNICEF. These two agencies were created after World War II to address urgent problems of refugees and degraded conditions.

WHO has created the concept of the "water service ladder" to describe drinking water access across populations. It has rungs for "unimproved" and "improved" water sources. "Piped water on the premises" is a separate category, but is often lumped into the approved source category. These classifications relate more to low-income countries than to countries with reliable water supply utilities and near-universal access to water supply. JMP assumes that if the water source is on the list of improved sources it "… adequately protects the source from outside contamination…" and will meet WHO guidelines or national standards (Joint Monitoring Programme 2022).

Examples of improved sources include piped water into a dwelling or to a yard or plot, a public tap or standpipe, a tubewell or borehole, a protected dug well or spring, and rainwater. Unimproved source examples are unprotected springs or dug wells, a cart with a small tank, a tanker-truck, surface water, or bottled water.

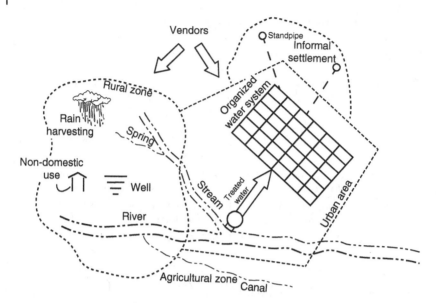

Figure 8.3 Community water supply showing access to sources.

To depict these sources, Figure 8.3 illustrates a fictitious community with an organized water system receiving treated water and distributed to some residents. Water is also provided by vendors and standpipes for other residents. Still other residents live in an informal settlement, which may lack reliable access or any access at all to water supply. Sources of water in the rural zone can include a stream, a spring, rainfall harvesting, or a well. The different classes of service illustrate inequities in providing safe water to all residents, a problem of environmental justice that is explained in Chapter 16.

The responses by water managers to threats to drinking water can be illustrated by Figure 8.4, which shows the pathway from source through treatment, distribution, and plumbing systems. Various threats due to contamination, failure, natural disasters, and other causes threaten each of the elements along the path. The consumer may obtain water through the piped system, through self-supply, or through a vendor. Vulnerability and responses to the threats depend on which water supply system is utilized.

Environmental Integrity

Water quality and public health are closely related to environmental integrity, which is a concept that means the general condition of the natural environment and its capacity to serve its intended purposes. Analysis of environmental

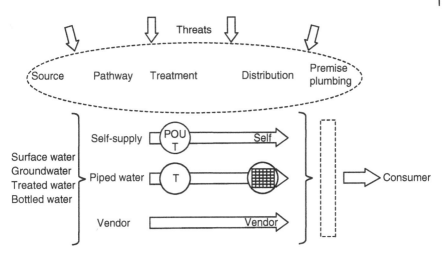

Figure 8.4 Threats to water supply and pathways to consumers.

integrity in water resources management is facilitated by ecohydrology (discussed in Chapter 5), which involves water quality because plants, animals, and ecological systems require given amounts of water of acceptable quality delivered at the times needed. Examples of ecohydrology in water management that address environmental integrity are to compute the consumptive uses of plants and their water quality needs and the needed instream flows for fish spawning and growth. Given the many types of ecological zones, plants, and fish species, this obviously is a complex area for assessment. One avenue toward assessing them is to study the condition of water resources in the United States.

The condition of water measures the readiness of a nation's water resources to meet demands placed on them. It is indicated largely by the condition of its surface waters, which are used and reused for many purposes as they pass through the hydrologic cycle. These many uses affect environmental integrity, so the indicated goal is to maintain the waters in good condition as measured by water needs of natural systems. Environmental integrity is challenged by conflicting demands, lack of consensus about what comprises good condition, complexity in assessing the condition of water, and in many cases by poor governance in setting and maintaining water standards.

The overall condition of surface waters is a function of water availability and quality, and these depend on natural conditions and human activities. Water availability in the United States is managed mostly through the water right systems of state governments, whereas water quality is regulated mostly under the federal Clean Water Act. One of the lessons from the Act is that assessing the condition of waters to measure trends and inform policy is complex and difficult.

Water resources assessment techniques have evolved for several decades. National water assessment was required under the 1965 Water Resources Planning Act, which instructed the Water Resources Council to prepare national water assessments. For an international focus, the focus on water assessment evolved with recognition of emerging water problems and was emphasized at the 1992 United Nations Conference on Environment and Development, which called for a holistic definition of assessment to encourage nations to develop data management for socio-economic and environmental purposes.

Despite the attention to it, terms used to explain the assessment of condition of water lack standardized definitions. However, the framework shown in Figure 8.5 can be used to explain the process and the terms. Assessment is the process used to determine condition, which is a measure of the status of water relative to its intended uses, considering both availability and quality of water for the uses shown.

The term availability requires measures of water quantity and timing of flows for natural systems, such as to supply water for fisheries. The term adequacy signals that while availability and quality are essential metrics, an integration is also needed as an indicator of environmental integrity. As no single measure can represent all attributes of water condition, a case-by-case basis for regions and scales of systems is needed.

While the United States does not conduct a unified and formal national water assessment, it has a piecemeal assessment approach that traces back to the initiation of programs to gage rivers and streams by the US Geological Survey and to its

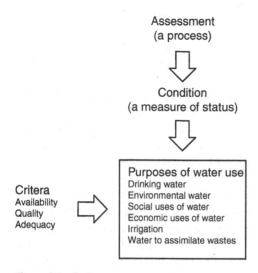

Figure 8.5 Definitions relating to water resources assessment.

predecessors. The US Army Corps of Engineers developed water resources assessments to support its river basin plans and policy analysis and after its creation, and the Bureau of Reclamation also began basin assessments to support development projects. Now, USEPA's legislation provides it with a patchwork of assessment authorities across the environmental media.

The concept of ecosystem services provides us with a way to explain the importance of environmental integrity and how natural ecosystems meet our needs. They are generally explained in terms of four general types: provisioning services (like water, food, and other resources); regulating services (like water and air quality improvement); habitat or supporting services (like habitat for species); and cultural resources (like health and recreation, as well as aesthetics). Clearly, effective water management is essential to sustain these vital services.

References

Eisenberg, J.N.S., Bartram, J., and Hunter, P.R. (2001). A public health perspective for establishing water-related guidelines and standards. In: *Water Quality: Guidelines, Standards and Health: Assessment of Risk and Risk Management for Water-Related Infectious Disease*. WHO 2001 (ed. L. Fewtrell and J. Bartram), Chapter 11. IWA Publishing http://www.who.int/water_sanitation_health/dwq/iwachap11.pdf.

Joint Monitoring Programme (2022). Progress on household drinking water, sanitation and hygiene. https://washdata.org/report/jmp-2021-wash-households (accessed 5 May 2022).

Questions

1 Give two examples each of physical, chemical, and biological water parameters.

2 Identify two water quality parameters that can be measured in real time in streams or wells.

3 Give two examples of how runoff can change the mineral content of streams.

4 What are three main sources of environmental water impairment according to EPA?

5 What happens to lake water quality when substantial evaporation occurs?

6 Make a diagram to explain the nitrogen cycle.

7 What happened after the advent of filtration and chlorination to cause continuing water quality problems?

8 Explain the likelihoods that people in rural areas, cities, and informal settlements have access to safe water.

9 Explain the nineteenth century discoveries that explained the link between biological contamination and disease.

10 Explain how the multiple barriers of water protection relate to probability of exposure in risk analysis.

11 Why does the JMP use the term "improved sources" for access to water

12 What is meant by public health, how does it differ from medicine, and what is the role of engineering in it?

13 For each of the following categories of contaminants identify a likely type of disease: Giardia, disinfection byproducts, arsenic, copper, lead, nitrate, benzine, and PCBs.

14 What is the difference between acute and chronic disease?

15 If a person becomes ill with diarrhea, what is the likely water-related pathway to the illness? What would be the likely water-related pathway to a skin infection?

16 What are the WHO and the Joint Monitoring Programme (JMP)?

17 Define environmental integrity and explain how it relates to water quality and public health.

18 Define ecosystem services and explain how they relate to water resources management.

19 Give examples of water-related ecosystem services in the following categories: provisioning, regulating, habitat, and cultural resources.

20 How do the concepts of One Health, One Water, and Environmental Health relate to water resources management?

21 For each SDG goal, identify one way that it links to water management.

22 How does the water resources assessment process incorporate water quality, public health, and environmental integrity?

23 Identify how drinking water service, wastewater service, and irrigation might cause diseases and the types of disease.

24 Explain the key functions of water managers that affect waterborne disease and public health.

9

Models and Data for Decision Making

Introduction

Water managers rely on data and models to inform their decisions, just as managers in business and government do. Data analytics and modeling have evolved rapidly, and they are no longer specialty tools but are now commonly used in planning, design, and operations. Data science provides broad insights about the possibility of discovering knowledge from data, and data analytics provides ways to apply math to data to answer questions and to make informed decisions. To support these trends, more water and satellite-based datasets are becoming available continually.

This chapter explains how models and data are used to support water management decisions. It describes what models do and their uses, data types and sources, model types and examples, and decision support systems. There are so many models available that it can be confusing to know which one to use for a given situation. Moreover, many firms have been organized to develop and market models, so the water manager must know how to choose among them to select the best one for the needs at hand. Once models have been obtained, running them effectively and maintaining them become issues.

Today's access to data and use of models builds on experience with water management from early civilizations, but there was not much data then. Exceptions are records for a few major rivers such as the Nile and historical writing about other water phenomena, such as about the water supply of Rome. Despite lack of data, water management principles evolved through empiricism and experience. Engineers and managers produced sophisticated reports based on limited data and computational capacity. The technical basis for water management was growing with theories about hydraulics and hydrology, such as development of

Water Resources Management: Principles, Methods, and Tools, First Edition. Neil Grigg.
© 2023 John Wiley & Sons, Inc. Published 2023 by John Wiley & Sons, Inc.

equations for flow and runoff. These early theories provided much of the technical basis for the models we use today. Empiricism is still used and has become more powerful with greater access to data. Early empiricists were harbingers of today's analysts who use data in powerful and insightful ways.

Beginning late in the nineteenth century, data collection programs such as stream gaging were started. By the 1950s, flow and precipitation data were available for many streams and watersheds. In that decade, researchers in a water program at Harvard laid a foundation for use of models in water management. Advances led to computer-based models like the Stanford Watershed Model and the predecessors of today's hydraulics and hydrology software for runoff and flow in channels. The Harvard Water Program had pioneered use of models for decision analysis, which led to methods for water resources systems analysis and later a broader concept of water resources management.

During the 1960s through about the 1980s, emphasis was on constructing models, but the focus has shifted from model development to use of off-the-shelf model-based tools. With increasing access to data, how real-world water systems behave becomes clearer. Government agencies have improved their data access arrangements, and water managers can access data needed for many applications. Many places still lack hydrologic data, but gaps often can be filled by remote sensing data, including from satellites. In some cases, raw data are used directly, but statistics based on data may be required. Models require data for calibration and validation.

Model Uses for Water Resources Management

Models provide synthetic information to help us analyze problems, formulate, and test plans and solutions, and then to make decisions about system and infrastructure configurations and designs. They predict responses of real-world systems to forcing data like weather events and water demands, as well as many others. They are used in planning and engineering work, in control of systems, and in dispute resolution, including in court trials.

Models are useful in seeking integrated solutions by combining hydrological, technical, ecological, environmental, economic, social, institutional, and legal aspects to help find solutions to water problems. While hydrological models are well developed, those for ecological, environmental, economic, social, institutional, and legal analyses need much improvement, especially to incorporate stakeholder views. Integrated applications of models lead to use of advanced conceptual tools used in systems analysis and nexus concepts.

A water manager may use data and models to plan a water supply system by considering demand, supply scenarios, alternative systems, and the feasibility and risks of alternative plans. For an existing system, the manager may have a model-based decision support system (DSS) to provide real-time data and simulate results of decisions. For system operations, a DSS might take the form of a Supervisory Control and Data Acquisition (SCADA) system.

Different types of simulation and decision models are used in water planning and management. Simulation models explain how a watershed or a process responds and behaves. Decision models seek optimum paths and desirable trade-offs. Both types are part of a universe of models used for management of many types of natural, built, and social systems. The evolving discipline of systems engineering embraces modeling for many purposes through "model-based systems engineering" which, according to the International Council on Systems Engineering (INCOSE) is the "formalized application of modeling to support system requirements, design, analysis, verification and validation activities beginning in the conceptual design phase and continuing throughout development and later life cycle phases." Systems engineering captures a generalized view of modeling, and can be applied across fields, including to water systems.

As an example of use of a suite of models in a decision support system, Figure 9.1 shows a watershed with different types of models that could be employed. They begin with simulation of the inflow to the reservoir, and they include elements of water management in the stream, in the diversions, and in the water uses. Compiling a decision support system to consider so many moving parts will be complex, to say the least.

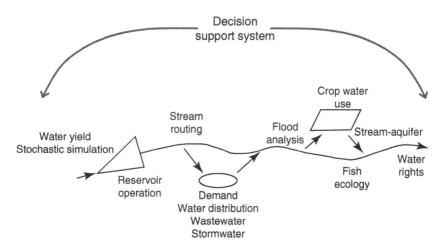

Figure 9.1 Models for analysis and to comprise a decision support system.

Structure and Functionality of Models

While there are many types of models, their underlying structure is normally about the same. They work by simulating how a system behaves when subjected to forcing data. The term forcing data is used widely in climate-related models, but it can be extended to any category of model. The forcing data become input data to be processed through a model engine, and then to generate output information. Some control function determines timing and quantity of the input signals. The engine predicts outputs from inputs. For example, you turn on a heater (control) to admit electric power to the heating elements, which generate heat according to a model engine that relates heat to power, and the result is heat for a room (output). In the case of a hydrologic model, the inputs are precipitation and other data. The model takes watershed characteristics such as size, slope, and vegetation into account to predict the outputs in the form of hydrographs and losses. The model produces hydrographs at selected points, and the user can vary inputs to study different scenarios. Models can incorporate feedback to compare predictions to observations and be improved by successive calibration steps through machine learning.

Watershed models illustrate how these processes work. Figure 9.2 is a block diagram of a simplified typical watershed model. The input is precipitation, which

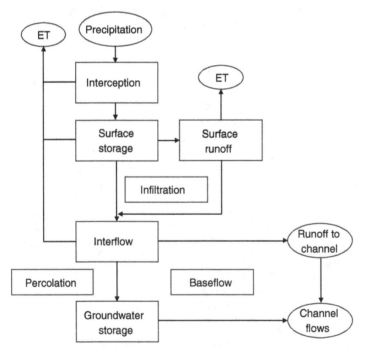

Figure 9.2 Block diagram of a typical hydrologic model.

is intercepted and then becomes surface storage, leading to infiltration. After infiltration and percolation, the interflow leads to groundwater storage, which affects channel flows.

Models like this can be organized with a user interface for data input, a database to drive the computations, a model engine to simulate the various interactions with equations and instructions, and an output format. For example, in the simple model shown in the figure, data can include the watershed characteristics as well as the forcing functions.

Data Types and Sources

Data needed to analyze behavior of water systems come from different sources. Land use and topographic data are needed to know the status of watersheds. Climate, weather, and snowpack data provide the drivers to analyze runoff and water availability. Surface and groundwater data indicate quantity, quality, and rates of change of water properties.

Water quality data indicate health of water in the environment and in water supply systems. Environmental data tell us about the health of plant and animal life. Other data indicate demographic changes, flood damage records, drought histories, use of infrastructure, and the performance of management systems like operations and finance.

Sometimes, data can be used directly to analyze water systems. For example, you can plot gage data on probability paper to establish flood frequency, extrapolate historical water use to predict future use, and analyze a stream's stage-discharge rating curve based on hydraulic measurements.

The main sources of water data are federal agencies. The US Geological Survey (USGS) is the main data source for stream flows and related hydrologic data, while the National Weather Service is the main source for climatic and weather data. The Natural Resources Conservation Service (NRCS) provides snowpack data. The USEPA has data on drinking water and water quality issues, and other federal agencies may be sources of specialty data, such as FEMA information on flood losses, USACE data on dams and river infrastructure, USBR data on western water systems, Fish and Wildlife Service environmental data, and USDA data on irrigation. Other sources of data include state and local agencies, especially utilities and special districts.

Data and model results are used to formulate performance indicators (PIs), which are indicators of progress toward an intended result and provide focus for improvement. PIs can be used for monitoring and provide an analytical basis for decision making by focusing attention on what matters most. They can be assembled into clusters, such as scorecards. Performance indicators should compress information, be precisely defined, and limited in number.

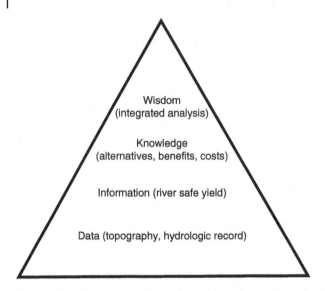

Figure 9.3 Information and knowledge hierarchy.

Figure 9.3 shows how river and topographic data can be processed to yield decision information on whether it is a good idea or not to build a reservoir. The raw data become information, which is converted to knowledge by analysis and processing. This in turn should become indicative of the best choice about how to proceed with the project.

There are many new data sources, including satellite data. With such increasing data availability, the emerging science of data analytics will help water managers to find patterns in records promises to transform water management. Using this more abundant data together with synthetic data using artificial or generated data sets, actual observed data can be augmented to reveal more useful decision information.

Types of Simulation Models

Simulation is the main engine of models used in water management. Its power is increasing, along with applications and functionality. This is a natural development, given the rise in computing power and need to test how systems behave without waiting to collect long-term data. There is competition among researchers and organizations to develop and improve models, both for scientific and commercial purposes. Most models can be purchased for reasonable prices, especially when compared to the workforce cost of learning and implementing them.

Different models are available in each category, but often a single model will be most popular. Reviews of the models are available from different sources around the world, with guides published by hydrologic agencies like USGS and its counterparts in other nations. USGS has been in the model development business for long time, and has developed and continues to support a number of specialized models for runoff, groundwater analysis, and other purposes.

The USACE's Hydrologic Engineering Center (HEC) is also a long-standing focal point of excellence in model development. USEPA has developed several long-standing models and continues to sponsor development of specialized packages, which are explained in different websites. The Agricultural Research Service (ARS) is the sponsor of the popular Soil and Water Assessment Tool (SWAT) model. An Internet search of the names of the models will normally take you to the current source of information about them, as well as download possibilities.

Many of the publicly available models also have commercial versions, and these are easy to find through Internet searches. Vendors of these models offer service and support, and they may attach useful interfaces that facilitate data entry and analysis.

Hydrologic and hydraulic modeling tools are used in most of the model types. Runoff and channel flow models are commonly referred to as H&H models, meaning hydrology and hydraulics. Process models can combine several other model types and be used to construct decision support systems. Many examples of these are based on the systems dynamics framework, which was developed back in the 1960s. Examples are Stella and Powersim. Geographical information systems (GIS) are not simulation models themselves, but they provide a framework for inclusion of spatial features in software packages that include models.

Examples of simulation models align with different parts of water systems that interact to determine outcomes in water management. Use of model types for water management scenarios are:

- Watershed runoff for flood analysis
- Channel routing for flood routing and channel capacity
- Reservoir routing for real-time operations
- Water quality for regulatory decisions
- Distribution systems to manage flows and pressures
- Groundwater for water rights and wellfield analysis
- Stormwater for system planning, analysis, and design

Table 9.1 lists common categories and popular models available through public sources. Most models are in continuing evolution, and water managers should consult the latest websites of the agencies involved. The websites shown may not be valid going forward, but the model frameworks have been supported for years and are likely to remain so.

Table 9.1 Publicly available models used in water resources management.

Category	Uses	Publicly available models
Hydrologic	Runoff models are among the oldest and best established of model categories. Used to predict runoff quantity and timing.	HEC-HMS
Hydraulic	Used to simulate dynamics of channel flow to include depth, velocity, and time variation.	HEC-RAS
Groundwater	Used to simulate the effects of well pumping, climatic conditions, and flow-through aquifers.	USGS MODFLOW
Pipeline networks	Used to simulate flow and pressure in distribution systems.	EPAnet
Water quality	Used to simulate water quality changes in watersheds and stream systems.	USGS sparrow EPA HSPF EPA QUAL2K ARS SWAT
Stormwater	Hydrologic–hydraulic–WQ	EPA SWMM
Reservoir routing	Hydrologic routing through reservoirs	HEC-ResSim

Models to generate weather conditions are evolving but have not reached the level of the other categories. Custom weather models can be coupled with runoff models to predict stream flows. There are additional specialty model categories, such as for dam break, water hammer, water demand, and various types of management models.

Capabilities of the models continue to expand. For example, river channel models can be obtained for analysis in one, two, or three dimensions and to simulate unsteady flow. They can be used to analyze water delivery, floods, bridge scour, sedimentation, and more. Models for reservoirs in series can be used to study water allocation, flood control, and drought response. Of course, greater capability usually means greater complexity and cost in terms of time required.

The scope and capability of water models continue to expand, and evidence of this is NOAA's National Water Model (NWM). The goal is to enhance the agency's water flow and flood forecast capabilities and expand them to more locations. Data will expand beyond USGS gaging locations to around three million locations. The model combines networks of existing stream gages and atmospheric modeling with soil moisture, surface runoff, snow water equivalent, and other parameters. NOAA sees the NWM as part of emerging water intelligence products to help communities and industries in water management decisions. The model engine is a community-based WRF-Hydro framework developed by the National Center for Atmospheric Research (NCAR).

Another ambitious model framework is HAZUS, which was described in Chapter 7. HAZUS is a GIS-based analysis tool distributed by FEMA to provide rapid, area-wide loss estimates for multiple natural hazards, including flooding. There is a HAZUS loss library, which contains a collection of risk assessments and data on economic losses, building damage and social impacts from historic events, and planning scenarios.

Another type of complex model is the "digital twin," which is a term to explain the concept of modeling a physical system with the computer simulation that enables the testing of behaviors and outputs that stem from different management actions. These are convenient to model behavior of a water distribution system or a system of water infrastructures comprising a reservoir and various water uses in a basin. Figure 9.4 illustrates the concept showing how the DSS is used to inform controls of various types of water infrastructure systems. The digital twin will be embodied in the dashboard and logic box of the DSS.

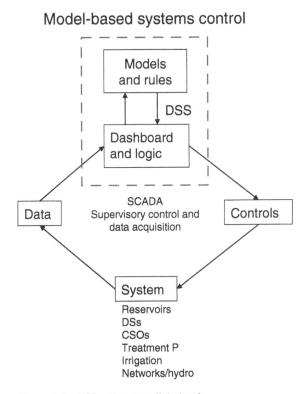

Model-based systems control

Models and rules

DSS

Dashboard and logic

SCADA
Supervisory control and
data acquisition

Data

Controls

System

Reservoirs
DSs
CSOs
Treatment P
Irrigation
Networks/hydro

Figure 9.4 DSS acting as a digital twin.

Example of a Hydrologic Simulation Model

The starting point of analysis in many water management situations requires hydrologic modeling, and to explain the concept at a basic level a simple example is provided here. The system to be analyzed is the one shown on Figure 9.1 and has a reservoir to provide water to a city, a farm, and instream flow requirements. The required capacity to be determined for the reservoir is to ensure no failure more often than once in 50 years in meeting the demands.

The main performance indicator is number of failures to meet the flow requirements in 50 years. Rules are based on meeting demands posed by the city, irrigation, and instream flows, as well as water quality. An additional constraint is that the channel flood capacity in reach CD is not to be exceeded in the same time period.

The study process involves developing the model, data collection, testing the model, scenarios, and analysis. After the model has been tested, you would create alternative solutions for reservoir capacity, run the model, and check performance indicators to find the solution that meets the criterion, usually with lowest total life-cycle cost, considering the TBL.

For simulation of the system, you need data on reservoir inflows, alternative reservoir capacities, areas, and evaporation. Normally, reservoir surface area and evaporation vary with water surface elevation, but to make this model simple we will use a constant reservoir surface area. You also need to know the system losses and the demands for city, irrigation, and instream flow needs. You should also know the flood conveyance capacity of the channel and priority of taking shortfalls.

The state variable is reservoir storage at any time. To build the simulation model, you begin with the fundamental equation of hydrology:

Change in storage in a time interval = inflow volume − outflow volume

Running the model involves computation of the water balance for each time interval and determining the change in the reservoir storage. You should establish a time step. For this example, we use a monthly time step.

To develop the model for the reservoir, DS = Q_A − Q_B − Evap; where DS = change in storage in the reservoir in a given month. Q_A = monthly inflow and Q_B = release in the same month. The value of Q_B determines whether needs can be met in the reaches below the reservoir.

From one month to the next, ResLevel$_t$ = ResLevel$_{t-1}$ + DS where ResLevel$_t$ and ResLevel$_{t-1}$ are reservoir storage quantities for month t and the previous month, $t-1$.

For the overall system, Q_G = Q_B − DivC + RetD − DivE + RetF − syslosses. This considers all diversions and returns, and Q_G is the ISF in reach F–G.

These equations are for the general situation, but the flow in reach C–D must be checked for ISF compliance because some water has been diverted from this reach. Flood checking in CD is like an upper bound on water levels in that reach. This was omitted from the example to keep it simple, but it shows how, to the extent possible, you must maintain flows in CD above the minimum ISF and below the flood level.

For this simple example, we have the inflow data and do not require calibration with historical records. However, if you included data such as losses internal to the system, you would require a calibration and validation process to identify and quantify the parameters.

To run the model, we simulate inflows with actual or synthetic data that forces the system to respond. If you have a long record, you could run it through the model. However, even if you had 50-years, that would only be one sample of many possible 50-year periods. For this reason, generating synthetic data is used here to illustrate the concept. You can generate as many years of synthetic data as you like to extend the analysis.

A limited set of synthetic data for sets of 10-year inflows has been developed for this example. You could also consider random, weather-induced changes or gradual climate change in generating the synthetic data. Statistical approaches to generation of synthetic data often use AR (auto-regressive) or ARMA (auto-regressive, moving average) models. With these, you create a synthetic time series that preserves the statistical attributes of the data you have. Synthetic data are not real data but provide statistical fluctuations that enable us to study how our system will behave under various possibilities, given a probability distribution.

For this example, our simple model uses a constant distribution of monthly flows, which we impose on a series of annual flows. We vary the yearly flows to show wet and dry periods. The monthly flow distribution and a 10-year series of annual flows are shown in these tables. The average for the 10 years is $Q_m = 212\,973\,\text{AF}$.

Month	% of annual inflow	Year	Total annual inflow (AF)
(a)	(b)	(c)	(d)
Jan	0.030	1	183 980
Feb	0.045	2	161 250
Mar	0.100	3	193 500
Apr	0.129	4	268 750
May	0.174	5	301 000
Jun	0.149	6	290 250
Jul	0.104	7	247 250
Aug	0.082	8	182 750
Sep	0.068	9	172 000
Oct	0.050	10	129 000
Nov	0.039		
Dec	0.030		

To use the 10-year average and generate a series of annual flows, we let each flow depend on the past year only (lag-1 autocorrelation). We first provide a deterministic component based on a lag-1 autocorrelation coefficient:

$$Q_i = Q_m + \rho^* \left(Q_{i-1} - Q_m \right),$$

where, Q_i is annual flow, year i, Q_m is mean annual, ρ is lag-1 autocorrelation coefficient, and Q_{i-1} = annual flow, year $i-1$.

The random component is usually associated with a given probability distribution and the variance of the historical flow record. We use a simple approach with a uniform distribution and a simple ratio of the mean flow as a measure of the variance. Let the random component e be:

$$e = 2^* (\text{RAND}(\ ) - 0.5)^* Q_m / 3$$

The quantity 0.5 is subtracted from RAND to enable a range of -1 to $+1$ when multiplied by 2. RAND() is the random number generator function in Excel. We divide by 3 to set the range of the random fluctuations at 1/3 of the mean flow. This is an arbitrary decision meant to generate a desired range of variability in the analysis. You could sample from a given probability distribution rather than the uniform distribution to obtain more variability in generated data

The resulting synthetic flow generator with the random component is:

$$. \ Q_i = Q_m + 0.3^* (Q_{i-1} - Q_m) + \text{RAND}(\ )^* Q_m / 3$$

where we take ρ is 0.3 for the auto correlation coefficient. This can be changed to consider more possibilities.

Using this equation, we can generate a series of any length. How long the series should be depends on the study goal. For example, if you check 100-year performance, you need to check multiple 100-year records. This could be done by checking a record of 100-year traces, say over 2000 years of flows. This would yield 24 000 data points, making Excel unwieldly, so you would probably use a different computational method.

To illustrate a simple process, we use 20 years of monthly data, or 240 data points, and route the data series through the reservoir. A simplified demand schedule is used, which is shown here as demands from the reservoir in cfs:

Jan	126
Feb	128
Mar	129
Apr	140
May	329
Jun	519
Jul	536
Aug	511

Sep	485
Oct	290
Nov	127
Dec	126

Two runs are made (Figure 9.5). One assumes a 50 000 AF reservoir, which does not perform adequately and is unable to meet the demands (graph on the left). Another run (graph on the right) shows that a 150 000 AF reservoir meets the demands without running out, but it comes close to running out around month 100.

(a)

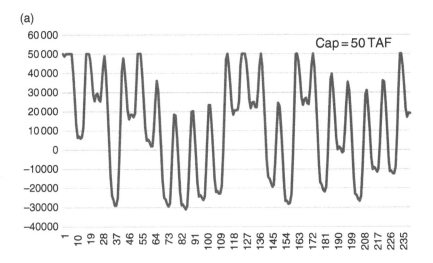

(b)

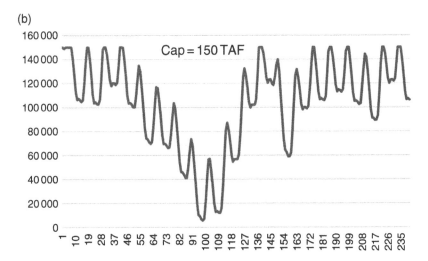

Figure 9.5 Reservoir performance for (a) 50 TAF and (b) 150 TAF.

In both cases, the reservoir was assumed full in the first month, but this assumption can be changed. These graphs are based on only one run each and represent only one set of possibilities of the randomly generated inflow sequence. While many assumptions were made to perform the analysis, the basic process can be used to estimate the size of reservoir needed for the conditions specified.

Planning Models and Decision Support Systems

Planning and decision models are widely used in water management. Planning models focus on evaluation of alternative problem-solving scenarios, and decision analysis is a systematic way to evaluate the pros and cons of different decisions. These must be crafted for individual situations, and they often use a commercial background program or shell. The concept of the decision support system (DSS) can be viewed as a planning and decision model, even as it may contain other models and data systems. Planning and decision models can be used for different management purposes, including shared vision modeling.

A DSS typically includes a user interface for data input, a model engine, a database, equations, computer instructions, and an output format. Various features can be added to a DSS, including PIs, additional models, mapping, and rules for logic and decisions.

When models and data are integrated, they become like a DSS. A good example is the Water Evaluation and Planning (WEAP) model system, which has evolved for more than three decades as planning tool for evaluating water demand and supply patterns and exploring alternative scenarios. The software is now managed by the Stockholm Environment Institute.

A DSS can be used in conjunction with a multi-criteria decision analysis (MCDA) framework to examine alternative scenarios. A MCDA set up is not a model itself, but it provides a framework to evaluate information, consider different categories of goals and tradeoffs. It stems from a concept from public sector economics called utility theory. In water management decision analysis that considers economic/financial, social, and environmental goals, the TBL and MCDA are useful tools. They are discussed in Chapter 13.

Optimization models can be included in a DSS to find the best value of a dependent variable as independent variables are varied according to different scenarios. As software tools were developed, different methods such as linear and dynamic programming and various heuristic methods have been developed. The Excel software has an optimization feature now. The fields of simulation and optimization continue to expand and you can follow their development through the Decision Analysis Society of INFORMS, which is the acronym for The Institute for Operations Research and the Management Sciences that was formerly known as TIMS and has also absorbed the Operations Research Society of America.

Questions

1 Outline the types of models and data that would be required to analyze whether a wastewater treatment plant could meet downstream water quality requirements under any hydrologic scenario.

2 Explain how the disciplines of water resources systems analysis and systems engineering compare with each other and differ.

3 Define empiricism and give an example in water management.

4 Explain what is meant by the phrase "data analytics"?

5 Name five types of water resources data used in analysis and management and indicate their principal sources.

6 Identify a common water resources decision scenario and explain the need for data and models to confront it.

7 Explain what a performance indicator is and give one example.

8 Identify a performance indicator that requires diverse types of data in support of a water resources decision and show how the data would be combined.

9 Draw a block diagram to indicate the elements of a generic simulation model. Expand the diagram to show how it indicates the components of a typical watershed model.

10 State your opinion about the most influential developments in computing and modeling, give dates that they occurred and an example of how they changed the practice of water resources management.

11 Explain why, how, and when watershed modeling first began.

12 Sketch a river channel and illustrate one-, two-, and three-dimensional flow, and steady and unsteady flow.

13 Sketch a profile of two reservoirs in series and make notes to illustrate how a reservoir simulation model would work.

14 Explain the features of the SWMM model and give examples of how it would be used.

15 Explain the general features of the MODFLOW groundwater model.

16 Draw a simple pipe network with two or more loops and make notes to illustrate the principal components simulated by the EPAnet model.

17 Sketch a block diagram to illustrate how a decision support system will work.

18 Select a type of model and give an example of its use in planning.

19 What is shared vision modeling and what is it used for.

10

Operations, Maintenance, and Asset Management

Introduction

Once water facilities are built, they require effective operations, maintenance, and asset management to sustain good performance, manage risk, and extend their service lives.

The systems to be operated are those explained in Chapter 3: reservoirs, water sources, treatment plants, distribution and wastewater collection systems, stormwater systems, irrigation systems, and aquifers. Each has its own configuration and operating rules.

While operations and maintenance are often lumped together as "O&M," they involve different management processes. Sometimes they are put under "operations," and some managers advocate replacing "maintenance" by "management" to indicate that more than preventive maintenance is involved. In many ways, this need is met by the process of asset management. Ultimately, the process of operations aims to control the systems to achieve performance targets, maintenance involves prevention of problems and performing minor repairs, and asset management is a general process for tracking and renewing infrastructure components as needed. Major repairs can be considered as renewal, rather than maintenance, but there is often no sharp dividing line between them.

Operations, maintenance, asset management, and system renewal are parts of a life-cycle continuum of using and caring for infrastructure. These are essential for sustainability, good performance, and accountability in using public resources. The facility lifecycle is shown on Figure 10.1, which illustrates an ongoing O&M phase and a renewal phase that occurs in the capital cycle and happens periodically. Lifecycle management means essentially to minimize all costs by working appropriately on each phase of the cycle.

Water Resources Management: Principles, Methods, and Tools, First Edition. Neil Grigg.
© 2023 John Wiley & Sons, Inc. Published 2023 by John Wiley & Sons, Inc.

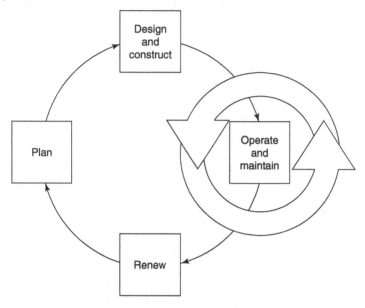

Figure 10.1 Maintenance and renewal in the facility lifecycle.

This chapter explains the theories of operations, maintenance, and asset management and how, with effective planning, design, and construction, they can assure the integrity of water infrastructure.

Organization for O&M and Asset Management

O&M may be combined in an organizational unit (Figure 10.2), but asset management is usually a separate staff function, often under engineering but maybe placed elsewhere, such as in planning or related to finance or data management or even bundled with data management and GIS. A simple organizational scheme is shown to illustrate these functions and how they might be placed within a water organization. Sections such as the laboratory, along with staff functions such as legal and accounting, are not shown for simplicity. Asset management and the GIS section usually involve different parts of the organization and lines of cooperation could be shown to illustrate work across sections, such as data management and GIS in support of planning, engineering, and operations.

Operations Management

As a general concept, operations can refer to all management activities, as in running all aspects of a water utility. When applied to a water infrastructure system, it means to exercise controls such as opening gates or actuating pumps, to monitor

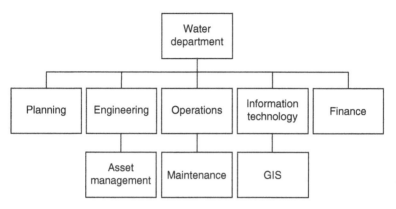

Figure 10.2 A basic water department organization.

performance, and to adjust according to feedback received. It may use a Supervisory Control and Data Acquisition (SCADA) system for monitoring and controls. The SCADA will operate according to the operational plan and under the constraints and guidelines that are imposed.

SCADA technologies started to evolve before the advent of computers, and their improvements run parallel to those for computing and new technologies for instrumentation and control. SCADAs are widely used in process control systems of different kinds, and their use can be extended to water resources systems. With advances in instrumentation and telemetry, movement toward automatic operations continues to advance. The risk of this, of course, is potential vulnerability to malevolent activity.

Figure 10.3 shows the general concept. A system such as a treatment plant handles inputs and provides outputs. It operates under and is monitored and possibly controlled by a SCADA system. Documentation such as an operating manual with guidelines will be available. The performance checking will produce feedback to adjust inputs and controls. On an overall basis, the operational tasks are informed by a discipline called control engineering.

Such SCADA-operated systems, together with automation comprise the basis of "smart" water systems, which are generally those with some built in intelligence. Another title is intelligent water systems.

At a large scale, operations could be of a major reservoir or a series of reservoirs, such as those run by the USACE. Many water and wastewater treatment plants and distribution and wastewater collection systems must also be operated. Irrigation systems also require operations, where the approaches will differ per the size and sophistication of systems.

Automation of control systems is practiced widely, and requires sensors, communication links, and actuators. In the field of control engineering,

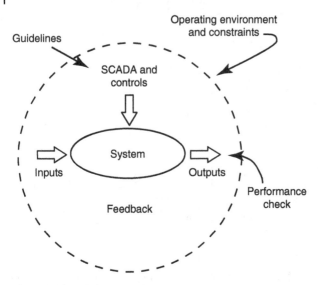

Figure 10.3 SCADA operating environment.

programmable logic controllers (PLCs) are widely used, and their concept can be extended to more sophisticated control systems at larger scales. Many vendors across industries supply components for this growing industry. Further information about process control is available by searching under that title.

System operations requires operators, and the title can be used in different ways. It generally refers to the person with main responsibility to make operational decisions. For water and wastewater systems, state governments have special certification programs for operators, and these are important in quality assurance of operating results. USEPA requires that states have these programs and will withhold funds if they do not. The Association of Boards of Certification, located in Iowa, provides information and support for several categories of water operators, including water treatment, wastewater treatment, distribution systems, collection systems, and others like industrial treatment and biosolids.

At a higher level, system operators may be responsible for complex operations of reservoirs or other interconnected water systems. They may use mathematical models and operate at a high engineering level as opposed to a level where they are manually following operational manuals. In any event, they will operate according to guidelines that are important and should be updated regularly. In the case of operating reservoirs, the USACE has published guidelines which can be downloaded by searching on reservoir operations.

Smart Water Systems

Operations of water systems leads naturally into the emerging concept of "smart" systems, which use computer controls and information to collect data, use that data to improve operations, and communicate with citizens about their lives in cities. Smart water systems can be in urban areas, for irrigation, or for any configuration of water infrastructures working together. The most frequent application is for smart water systems in smart cities, which could feature controls, collection of system and user information, application of the information to control water operations and to inform citizens, and for emergency management.

The concept has evolved from the early SCADA systems to today's concept. The main difference today is to include transmission of data shown by the dotted lines in the Figure 10.4 to provide information for operations, customers, and emergency management simultaneously.

These features evolved from development of SCADA systems, databases, and GIS to create technology-based platforms for system management. When

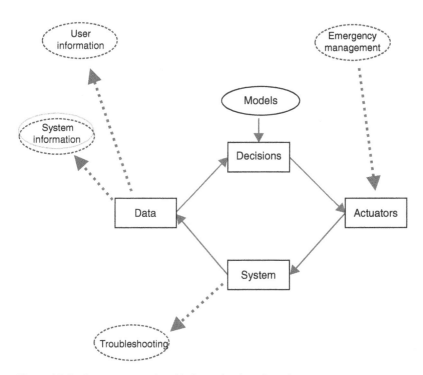

Figure 10.4 Systems control and information interfaces in a smart water system.

personal digital assistants became available, they were widely used in control and data acquisition functions. Currently, the technologies have advanced with development of smart phones, smart meters, and improved process controls. In urban areas, they are pointing toward more automation and management information for system tasks such as leak detection, cybersecurity, and to inform users.

While the interest in smart water systems has been driven by technologies and control methods, how customers use smart water information to help transform cities is more strategic. If the past is a guide to the future, it is expected that technologies and control methods will continue to advance incrementally, new business opportunities will be created, and experts and managers will continue to seek out new methods to experiment with them. Meanwhile, the benefits of these advances to customers and how the evolving smart systems are accepted and improve quality of life in cities remain larger questions.

Maintenance Management

Effective maintenance of water infrastructure systems is required to enable them to operate as intended and to sustain their condition. While maintenance might seem like a routine task, it involves important and sometimes-complex management procedures and water managers should take the lead in advocating and managing it. Otherwise, maintenance may be neglected as many facilities are out-of-sight and out-of-mind.

Maintenance includes preventive and corrective tasks. Preventive maintenance is a regular activity to enhance functionality and extend lives. It utilizes information sources like manuals, product information, and experience of workers. Record-keeping is important to include equipment data, the record of maintenance performed, repair history and spare parts records. Corrective maintenance for equipment or components can range from minor to major repair.

A MMS is an organized way to schedule, perform, and assess maintenance activities. It offers a framework to manage essential activities through a systems approach to preventive and corrective maintenance activities. One view of a MMS is as a sophisticated "to do list" with a preventive maintenance (PM) schedule and work order tracking.

Maintenance can be proactive and involve planning and deliberate attention to future needs, or it can be reactive and simply fix components when they fail. While the need for effective maintenance is obvious, it lacks immediate political benefits and leaders receive little credit for it. Regardless, its benefits can impress even a hardened financial officer or uninformed citizen, provided they are explained well. Unfortunately, maintenance benefits are often not explained effectively to managers, and boards, and the public may resist rate increases to pay for it making maintenance budgets are easy to cut. Therefore, even during financial shortfalls, it is important to make the case for the benefits of maintenance.

Asset Management

Water systems are often described as "capital intensive" because the value of their long-lasting physical assets is many times annual expenditures for operations. Managing these assets through their infrastructure life cycle can be challenging but applying data-centered management can help. The term asset management has been developed to express this concept and to explain how tools for lifecycle management can be coordinated.

Asset management is different from maintenance management in the sense that maintenance is an ongoing process to care for assets, whereas asset management is a high-level and enterprise-wide process to use all means available to extend lives and obtain peak performance from infrastructure systems.

The basic idea in asset management is to collect and manage substantial data on infrastructure systems and their components to assess condition and need for interventions, such as rehabilitation and renewal. A simple definition is that asset management is an information-based enterprise-wide process for lifecycle infrastructure management. The concept is illustrated in Figure 10.5.

Asset management concepts were developed to optimize the lifecycle value of physical assets while maintaining required service levels. Although the basic concepts are well-established, new methods, tools, and concepts continue to emerge for integration of asset utilization, performance, and business requirements.

Asset management systems are intra-organizational partnerships between the financial, and information sections with planning, engineering, operations, and maintenance. Water organizations have developed many formal programs to implement them, often with assistance from consultants. Software vendors offer off-the-shelf systems as well. The assignment of asset management tasks within an organization can be shown this way.

Organizational unit	Asset management activity
Budget and finance	Capital budget and accounts
Planning	Needs assessment
Engineering and construction	Capital improvement program (CIP)
O&M	Maintenance management system (MMS)
Information systems	GIS, databases, inventory

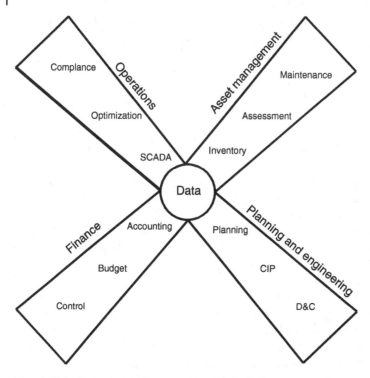

Figure 10.5 Data-centered management of physical assets.

The use of data from an asset management system might be for reporting, such as this hypothetical example: "The organization's physical assets are valued at $(insert value) million, and during the past year they delivered improved service over the previous year that was indicated by (evidence of better service and fewer failures). Investments in renewal were $(insert value) and improved condition assessment and repair methods have extended the lifecycles of existing assets by xx percent, representing a savings of $(insert value) for the city's utility and public works customers." As you see, the information provides a way to demonstrate how well an organization is using its capital assets to accomplish its mission.

There are many complicated explanations of asset management, but it reduces to a few key steps, such as these identified by USEPA (Figure 10.6):

In this simplified view, the starting point is condition assessment of current assets. This is important for all water infrastructure systems, but water distribution systems have come under scrutiny for being a particular challenge. A 2006 report by the National Research Council identified the importance of distribution system integrity on water quality, and the AWWA has published several reports about how the condition of distribution system is getting worse due to underinvestment. The

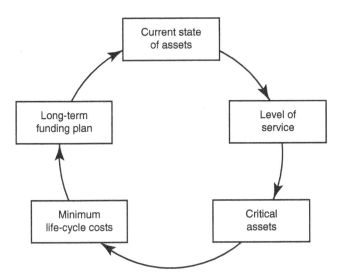

Figure 10.6 A basic five-step asset management process *Source:* U.S. Environmental Protection Agency / Public domain.

problem is a large inventory of legacy pipelines, many of which have already passed their design lifetimes. They are threatened by many external and internal threats, and renewal has not kept pace with deterioration. The primary reason for this is lack of funding and companion unwillingness of politicians to invest.

A good bit of attention has been given to sustainability of system integrity. The National Research Council report discussed it in terms of physical, hydraulic, and water quality integrity. Indicators included maintenance and corrosion control, management of main breaks, pressure management, biofilm management, and security measures, among others.

The next step is to determine level of service, so condition can be compared to that needed for required performance. Step three is identification of critical assets to ensure that they are cared for first. Next is the phase to determine the best way to manage the assets through minimizing the lifecycle cost. Finally, a long-term funding plan is developed. This is, of course, not a permanent solution, so the process is continuous.

Software for Maintenance and Asset Management

The glue that binds maintenance and asset management together is data. Maintenance management comprises activities for preventive maintenance, which requires scheduling, corrective maintenance, which requires work orders, inventory to provide facility data, such as capacity of equipment, date of installation, location, and others, and condition assessment as an ongoing process to make the organization aware of risk levels related to maintenance and renewal needs.

Inventory and condition assessment are related through data management because a facility's condition is a function of physical state, capacity, age, and other attributes that aid decisions about repair, replacement, and rehabilitation. Methods for condition assessment are evolving. Some are low-tech but others involve smart technologies involving new sensors and included as well nondestructive testing tools and methods.

Maintenance and asset management are also related through data to system planning, needs assessments, the operations budget, capital improvement program, and the capital budget. Figure 10.7 illustrates how data connect these seemingly diverse management systems. The maintenance management system requires the same data as are needed for system planning and needs assessments. When these are combined with financial information, a comprehensive capital management system can be organized.

Data are the basis for rapidly developing software systems for maintenance and asset management. A few decades ago, maintenance was often an ordinary activity comprising some preventive tasks and mostly hit-and-miss corrective maintenance on a reactive basis. Development of computers showed how data could be

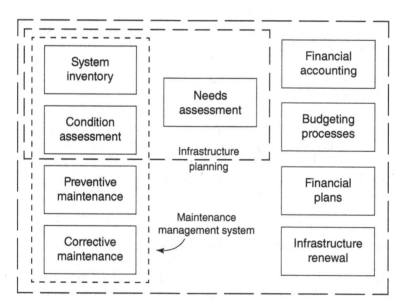

Figure 10.7 Data and information elements in asset management.

processed to yield valuable management information, and with the emergence of GIS, data bases and mapping could be combined to organize and direct work in new ways. Currently, water management organizations are acquiring software for inventory, asset management, and work order management to manage assets, track work, and improve financial management. Maintenance management software focuses on work orders and asset management software focuses more on financial issues, but some enterprise software systems can combine these. Regardless, the focus of asset management remains on finance and the focus of maintenance management is on system operations.

Applications to Water Systems

Dams should be monitored for safety, and operating components such as spillways and outlet structures require periodic inspection and repair. Reservoirs may require periodic cleaning or removal of sediment. At large scales, operation of reservoirs has major impacts on stream flow and the availability of water. In reservoirs, the main input is streamflow and the main output is releases from the reservoir. However, a reservoir water balance has other inputs and outputs such as precipitation and evaporation inputs, as well as seepage losses and groundwater inflow.

A system of water sources may include reservoirs, stream diversions, and wells. To operate these as a system requires knowledge of both the demand and supply. In the case of systems with many sources and demand spread over a large area, a model of demand and supply is required to determine an optimum operational plan. Such a model will show inputs, outputs, losses, and diversions to user systems. The SCADA could enable the operator to change sources to maintain a balance in the system.

The operation of a single well is relatively simple, compared to more complex water sources. However, when multiple wells with different constraints are involved, operation can include modeling and monitoring to ensure optimum and to avoid damage to the aquifer.

For water, wastewater, and stormwater systems, the operations section will normally manage both facility operations and maintenance to support goals for performance and compliance with regulations. The main facilities are treatment plants, piping systems, and in the case of wastewater, biosolids.

The operational plan for a water treatment plant will respond to demand, usually on an hourly basis, adapt to changing water quality, and monitor and report about water safety and the condition of the unit processes in the treatment plant. Such operation requires skilled and certified operators. The operational requirements for the wastewater treatment plant are similar in concept to water treatment plant. The unit processes are different, and in the case of the wastewater plant, more complex in chemistry and biology. AWWA sponsors a program called Partnership for Safe Water where utilities can help each other to improve treatment operations. The program has also been extended to distribution systems.

Water distribution systems are complex and can spread out over large areas. It is not unusual for a medium-size utility to have 1000 miles or more of piping. Operation of the system must take into account demand, water quality, security and more. The SCADA may be used to read meters remotely, operate district metering areas, utilize smart technologies such as an advance metering infrastructure, and to optimize the performance of the system.

Operation of wastewater collection systems is normally less complex than water distribution systems. Most wastewater collection systems operate by gravity and are simpler to operate than water distribution systems. Some systems include lift stations, storage, and other components. In such cases, a SCADA may be required to monitor the system. Monitoring for overflows and blockages is important in collection systems.

Much of the maintenance of water and wastewater is on the underground networks of distribution systems and collection systems, which comprise about two-thirds of system capital assets. Pipeline repair is a factor in controlling service disruption and cost.

Specialized guides for maintenance of water and wastewater systems are available, like from AWWA on system components, such as pumps and storage tanks. Given its heavy involvement in wastewater, USEPA has developed many publications on wastewater O&M.

While the systems are similar, their equipment and components vary in maintenance requirements. For example, water distribution pipes and wastewater sewers use different materials and design procedures. A treated water pump will be different from a sewage pumping system. A wastewater treatment plant uses different processes from a water treatment plant.

Examples of specialized water supply maintenance tasks include pipes, valves, pumps, hydrants, tanks, meters, generators, booster chlorination stations, and backflow prevention devices. Distribution system maintenance involves flushing to remove sediments, stale water, slime, and other unwanted constituents. Chlorine may be used to kill bacteriological growth. Cleaning may remove deposits in the pipe. Methods of cleaning include mechanical, air purging, swabbing, and use of pigs that are inserted into the pipelines to scrape and push the deposits and scales out. In older systems many valves may not have been exercised for years. In fact, some may not even be listed on records.

Corrosion is an important issue for distribution systems, and it might even impact health since metals leached from pipes such as lead can be harmful. Corrosion also causes financial damage such as staining clothes in washers.

Deposits can form from tuberculation, or formation of tubercles on pipe walls from corrosion. These roughen pipe walls, increase the "C factor," and increase energy required to pump water. Eventually flow can cease altogether. Pipe replacement may be required to cure the problem.

Maintenance (and operations) of water supply systems also involve monitoring to find cross connections. Also, the management of distribution systems may extend to providing guidance for maintenance of home plumbing systems, where the variety of situations is large and the range of materials is wide. Water auditing can be useful in distribution systems to identify needs for maintenance. AWWA has a management standard (G-200) that explains the most important processes of operations and maintenance of distribution systems.

Effective maintenance is critical for wastewater collection systems to avoid backups and associated problems. The Clean Water Act specified that new plants constructed through the construction grants program had to be preceded by sewer system evaluation and rehabilitation programs, complete with infiltration/inflow (I/I) analysis. This led to the use of new techniques such as video inspection to find problem areas. Methods for sewer system evaluation surveys include videos and other techniques to detect structural defects, corrosion, and problems like roots and blockages. Common maintenance problems associated with sewer systems include infiltration of groundwater, inflow of stormwater, clogging, breaks and damage from unauthorized and improper waste material.

Guidance for collection system maintenance is included in USEPA's Capacity, Management, Operations, and Maintenance (CMOM) program. CMOM was developed because the United States had invested many billions of dollars in construction of new wastewater plants, and they needed to be operated effectively, along with their associated wastewater piping systems.

Maintenance of underground stormwater pipes has received comparatively little attention, compared to water and wastewater. When a flood backup occurs, stormwater maintenance crews will fix it, but data such as historical breaks, regulatory sanctions, and performance measure are mostly missing from stormwater programs, mainly due to absence of regulatory oversight.

Although they both involve mostly gravity piping, maintenance of stormwater facilities differs from wastewater collection systems because many stormwater systems were managed jointly with street systems. Examples of routine work include vegetation mowing, trash, and debris cleanup, weed control, and revegetation.

Combined sewer systems are special cases of wastewater collection with the addition of stormwater and the possibility of periodic overflows. Monitoring and control of combined sewer systems involves a complex combination of wastewater and stormwater and the causes of overflows. In some cases combined sewer systems may include regulators, which may be placed inside the pipelines and require special approaches to control overflows.

Operation of irrigation systems begins with the source, such as diversion from a stream or pumping from a well, and continues through conveyance and distribution, even to small areas of farm fields. Diversion and conveyance to the farm field can involve losses, which should be monitored and controlled, to the extent possible. Once water is delivered to fields, and depending on the type of irrigation system, the irrigation can extend to individual plants or trees through smart operational systems.

Questions

1 What is the primary role of a system operator and what is "operator certification"?

2 What is a SCADA and how does it work?

3 Explain how the components of control systems (sensors, communication, actuators, PLCs) work in the context of a combined sewer overflow control system.

4 The concept of "smart" water systems is popular. What does it mean and how does it relate to system operations? What has been added to system management to provide more access to information for water users?

5 What is AWWA's Partnership for Safe Water and how does it work?

6 What does it mean that water systems are capital-intensive?

7 How does asset management differ from maintenance management?

8 List and explain briefly the five main steps in asset management.

9 Define "integrity" as it relates to water infrastructure systems. Give examples of metrics for it.

10 What do industry reports say about the condition of water distribution systems in the United States? What are the underlying factors? Are problems getting better or worse? Explain.

11 What is the difference between proactive and reactive maintenance management?

12 What factors might cause maintenance of water systems to be neglected?

13 List and explain briefly the four main processes of maintenance management.

14 Give examples of the tasks inherent in minor, preventive, and major maintenance.

15 How would a computerized maintenance management system work?

16 How do comprehensive

11

Water Governance and Institutions

Introduction

Effective water governance and institutional arrangements are critical to the success of water resources management. Water governance means the arrangements for oversight of water resources management and deals with policy, regulation, and support to enable utilities and public water organizations to perform their jobs. Institutional arrangements involve government structure and organization, laws and regulations, and other ways of life that has been established, like traditions and customs. These are like levers that control the water industry but are often out-of-sight and out of mind.

Water managers usually do not think about a concept like governance that seems abstract and academic, but the concept explains common statements like "water is mostly political," "water runs uphill to money," and "water engineering is easy, but institutional barriers block solutions." Such statements often appear when unforeseen problems occur or projects get stopped unexpectedly and people want to blame something out of their control or someone else.

To illustrate its importance globally, a key UN report stated that water governance was "… in a state of confusion" in many countries. It explained that "in some countries there is a total lack of water institutions, and others display fragmented institutional structures or conflicting decision-making structures." These realities blocked solutions to problems such as conflicting upstream and downstream water rights, diversion of public resources for personal gain, and poor application of laws and regulations (UNESCO 2006).

Definitions of water governance focus on elements such as political, social, economic, and administrative systems or rules, practices, and processes. These can seem fuzzy, but they mean how we organize ourselves through government to provide and control water resources management.

Water Resources Management: Principles, Methods, and Tools, First Edition. Neil Grigg.
© 2023 John Wiley & Sons, Inc. Published 2023 by John Wiley & Sons, Inc.

This is a short chapter to provide a framework for water governance and to set the stage for later chapters that explain its elements, like government organization and water laws.

How Water Governance Works

Water governance uses political, social, economic, and legal institutions to control water management. These include legislatures (political), financial structures (economic), partnerships (social), and regulatory agencies (administrative). These stem from the fundamental institutions of a state or country (like constitutions as legal institutions), and lead to organizations and processes, strategies, and outcomes, such as delivery of safe water.

Water governance operates through institutional arrangements, which mean social and legal mechanisms that have been "instituted." As Figure 11.1 shows, for water resources management they can be divided into formal institutions (at the top), like government actions, organizational roles, and laws and regulations, and informal institutions (at the bottom), like traditions and customs.

Formal arrangements begin with the rule of law and include the government, which was established in the United States in the 1780s. Many statutes for enabling actions and regulatory controls have been enacted through government authority since then. Some water organizations have been in place for a long time. Politics involve interest groups, who support movements such as to build or oppose dams.

Policy and strategy determine the high-level goals for water management. Policy is the highest-level water governance process because it establishes the framework for action and indicates what government intends to do based on public values and choices. Each water sector has its own policies and the full picture leads to "national water policy," although water policy cannot usually be fully integrated but must be a collection of sub-sector policies such as for drinking water or irrigation.

Regulatory controls comprise an important institutional arrangement in water management. They seek to find the right balance between individual and public interest. They are necessary because, at small scales people may work out issues, but at larger scales they cannot negotiate directly with each other and require formal arrangements to work the issues out.

Types of water regulations are:

- Health and safety (such as to supply safe drinking water)
- Water allocation (such as to recognize legal water rights)
- Environmental water quality (such as to maintain clean streams)

Fundamental institutions	Rule of Law Organized government
Institutional arrangements	Organizations, roles, processes
Institutional mechanisms	Government authority, legislation, regulations, compacts
Strategic decisions	Policy and strategy
Outcomes	Water systems management
Informal arrangements	Culture, incentives, networks, customs, behaviors

Figure 11.1   Institutional arrangements for water management.

- Fish and wildlife protection (such as to provide instream flows)
- Rates and charges (such as rates of a private water company)
- Service access and quality (such as adequate water pressure)

Regulatory processes begin with rulemaking by government agencies to implement statutory guidelines. Each type of regulation must have an authorizing statute and implementing agency. Enforcement requires mechanisms to respond to violations of regulations.

Informal arrangements begin with customs and behaviors that shape our attitudes and culture. The media exerts strong influences. Values and attitudes lead to behaviors such as building large houses that use a lot of water. An example of values that influence water management is the fishing industry, which needs clean water and good habitat. Also, some people are keen on conservation, while others want more development.

Some of the water management institutional responsibilities fall at the federal level and others at state and local levels. This is evidence of the principle of subsidiarity, which is a broad concept meaning that in social organizations, responsibilities should be undertaken at the lowest possible level. This is inherent in the US Constitution, which reserves powers not delegated to the United States for the states or the people.

Returning to the state of confusion in some countries about water governance, we can see that water management will be effective if the Rule of Law is in operation, there are well-designed water programs and regulations, adequate resources are provided for the public sector, citizens comply with rules, and effective public administration is practiced. If the Rule of Law does not operate well, officials are corrupt, citizens disregard law and order, and adequate resources and programs for water administration are not provided, the result will be failed water systems.

Water Governance and Management Compared

Water governance tasks span administrative and operational activities, and focus on empowerment, policy, oversight and regulation (Figure 11.2). Management executes policy to undertake the work of an organization. It depends on governance for policy (tell us what to do) empowerment (provide resources), oversight (monitor performance), and regulation (provide checks and balances).

Scale Factors in Water Governance

Water governance involves different arrangements that depend on the scale of the situation at hand (Figure 11.3). At small scales, like in a local community or neighborhood, the people and interest groups will work face-to-face on relatively simple

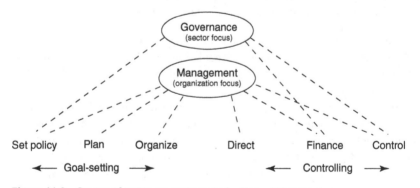

Figure 11.2 Scopes of water governance and water management.

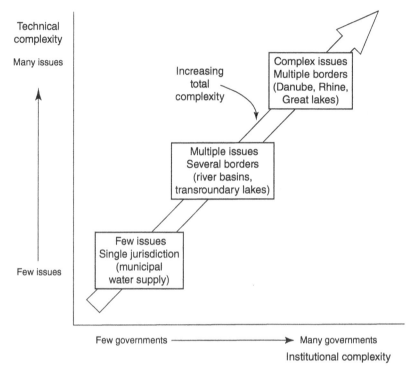

Figure 11.3 Complexity as a function of water management scale.

issues. The medium scale could include a watershed that crosses several political jurisdictions and involves several government entities with more complex issues. Still more complex issues are found at larger scales, which can include different cities, states, and even nations. Obviously, the larger the scale, the more difficult and complex will be the issues. A reality that arises from scale factors is that once a water management scenario moves from the smallest to largest scales, the actors have different relationships. At the neighborhood scale, they can meet in small groups and work out issues. At larger scales, they depend on representatives who have fewer stakes in the outcomes and may be replaced from time to time.

Shared Governance

The concept of shared governance means to distribute power to decide on actions. The term is used in different ways, such as the ongoing arrangements for shared corporate governance between labor and management or administration–faculty sharing of power in a university. In the context of water governance, the arrangements are more likely to be ad hoc and created to meet particular needs. The concept of "shared vision" modeling is intended to provide a tool to facilitate shared governance (see Chapter 9).

When the term shared governance refers to power-sharing across political units it could mean an inter-governmental agreement between levels of government or an inter-local agreement among cities and water districts. In any case, it is necessary to have a formal agreement whenever possible because successive political leaders might not honor informal agreements of the past.

Conclusions about Water Governance

Having key elements of water governance in place is essential to support effective water resources management. These include the rule of law, effective government, absence of corruption, effective regulation, transparency in public institutions, and demonstrated success in meeting public needs. Despite these being obvious, use of the term water governance seems abstract and difficult to understand. Governance is basically a basket of sociopolitical controls that determine the success of water resources management, after technical requirements have been met.

The chapters that follow address these sociopolitical controls by discussing water organizations, how water planning works, sociopolitical goals, economics and financing, and the important topic of how water laws and regulations are implemented in the highly regulated water industry. The overarching concept of integrated water resources management is presented in the final chapter.

Reference

UNESCO (2006). Water: a shared responsibility. Report2 of the world water assessment programme. http://www.unesco.org/new/en/natural-sciences/environment/water/wwap/wwdr/wwdr2-2006 (accessed 15 May 2022).

Questions

1 How does water governance relate to water management?

2 What are institutions and institutional arrangements?

3 Give three examples of institutional arrangements.

4 How does the rule of law affect water management?

5 Give examples of formal and informal institutional arrangements.

6 Considering the rule of law and formal water agencies, compare what you would expect in water governance in low- and high-income countries.

7 Why is it necessary to separate regulation and water delivery?

8 For the water sector, give examples of economic regulation and health regulation.

9 What does subsidiarity mean in the context of water governance?

10 Why is it necessary to separate regulation and water delivery?

11 For the water sector, give examples of economic regulation and health regulation.

12

Water Management Organizations

Introduction

In the model for water governance presented in the last chapter, water resources management organizations implement government authority and responsibilities through agency operations, regulatory controls, and intergovernmental cooperation. Water managers should know the identities and functions of these public and private water sector organizations to help them navigate types of situations with different players and objectives. Understanding water organizations and their functions will also help rising water professionals to advance their careers and to participate effectively in water organizations.

Because water management activities are fragmented, it is difficult but still useful to characterize the players as a coherent group that comprises an industry or a sector. The glue that holds it together is that water management addresses the single resource of water, although it appears practically everywhere in different forms and it meets a highly diverse set of needs. These attributes explain why water is more of a connector among sectors than a sector itself.

This chapter explains the framework of the water sector and the services it provides. Questions raised in the discussion are what types of organizations manage the water sector and its subsectors, what are their functions, and how do these functions interrelate to create a holistic approach to managing the water industry.

Framework of the Water Industry

Water management functions like an industry where the products are water-related services like drinking water, irrigation water, hydropower, and water for navigation. These are delivery mechanisms to provide water resources and respond to needs such as to operate utilities and infrastructure systems, provide flood control,

Water Resources Management: Principles, Methods, and Tools, First Edition. Neil Grigg.
© 2023 John Wiley & Sons, Inc. Published 2023 by John Wiley & Sons, Inc.

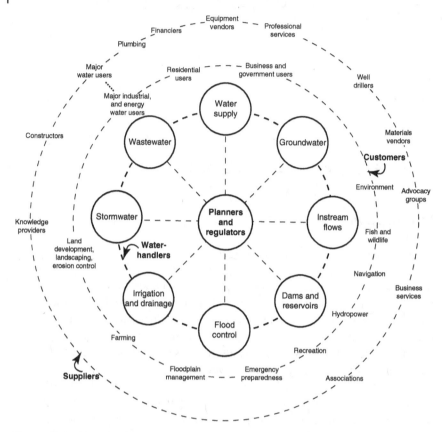

Figure 12.1   Map of water sector players.

and manage environmental water quality. These are overseen by regulators, who exercise control functions over water allocation, water quality management, and various other water industry activities. The service providers and regulators are supported by networks of support organizations that perform varied functions of planning, design, construction, equipment supply, training, and research.

Figure 12.1 shows a four-level view of this diversified water industry where utilities, planners, mission agencies, regulators, and supporting entities comprise the different types of organizational players. The first ring shows providers of water resources and services like water supply, wastewater, stormwater and flood control, and irrigation water. The term "water services" is not used universally, but it is a good way to lump utility-type public services together for discussion purposes. The term derives from the economic concept of "goods and services," where the services provide value but are not actually goods like products purchased at a store. The services provided in wastewater, stormwater,

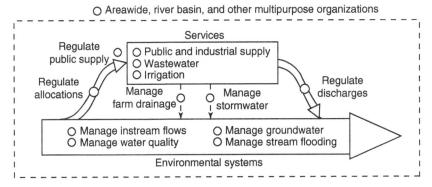

Figure 12.2 Management functions of water sector players.

irrigation, and flood control are delivery of performance, whether in terms of distributing water, collecting wastewater, or other.

It also shows water handlers who do not serve retail customers but manage dams and reservoirs, groundwater, and instream flows. The next ring shows the water use sectors, and the outer ring shows the support organizations, such as vendors of equipment and support services. At the core are the planners and regulators who provide much of the coordination of the water sector with activities that may not be very visible.

How these players operate water systems is shown in Figure 12.2, which shows water service providers, regulators, instream flow interests and organizations for areawide, river basin, and other multipurpose functions. Notice how the combination of the separate types of organizations with different authorities works to orchestrate a comprehensive approach to overall management. The policy design seems complete, but the overall effectiveness depends on how well responsibilities are carried out and, of course, how well individuals engage in conservation and small-scale resource management.

In the next section, these types of organizations will be explained, beginning with federal and state water agencies, then roles of areawide, basin, and other multipurpose water agencies will be explained. Organizations providing water-related services and their regulators will be described next, followed by a general support group such as vendors and water associations.

Federal and State Water Agencies

Early in the country's history, local and state governments were slow to develop capabilities, and the federal government was more active in water management than they were, which is the case today in other countries where governance remains weak. Examples of early federal programs in the United States include the USACE programs in navigation, organization of the USBR, and initiation of weather and stream gauging programs. Now, with increasing capability at the state and local levels, the principle of subsidiarity can be applied more effectively to have issues dealt with at lowest governance levels. The term is used in Europe to explain that unless it is necessary for the European Union to be involved, governance should be left to member states and in the United States, the same principle can be applied to devolve authority to state and local governments.

Over the past few decades, attempts have been made to characterize federal agency involvement in water management. Often, these attempts seek to find ways to increase efficiency and avoid overlap of functions and budgets. Despite good intentions, the "connector" attribute of water prevents it from being placed into only a few boxes. As a result, most attempts to characterize federal water activity conclude with long lists of water activities and organizations.

For example, a report by the Congressional Research Service (CRS 2017) provided a description of "Selected Federal Water Activities: Agencies, Authorities, and Congressional Committees." The CRS is one of the special-purpose agencies within the federal government that provide policy and legislative research and information to help Congress and the executive branch fulfill their missions. To explain their results, the CRS authors divided the discussion into four areas: water resources development, management, and use; water quality, protection, and restoration; water rights and allocation; and research and planning. They also discussed House and Senate committee jurisdictions over water resources and how numerous standing committees have jurisdiction over various components of federal water policy and how the wide range of federal executive branch responsibilities for water resources reflects these complexities.

Each of the four areas of discussion involves a long list of responsibilities. For example, the water resources development, management, and use category includes dams and dam safety, water supply development, groundwater, irrigation, water reuse, droughts, floods, hydropower, and more. These many fragmentary aspects of water resources management show up continually. As a result, and considering how water meets the needs of different sectors, a useful way to organize Federal water agencies is to view agencies by mission and how water plays into their activities. Figure 12.3 illustrates a simple concept for these.

The federal government does not have a single unified water department or ministry. In the figure, you can see nine cabinet-level federal agencies with major water programs. Under some of them is shown the main units that manage water, such as the USACE under the Department of Defense and the Bureau of Reclamation under the Department of Interior. Abbreviations that are used are shown in Appendix A.

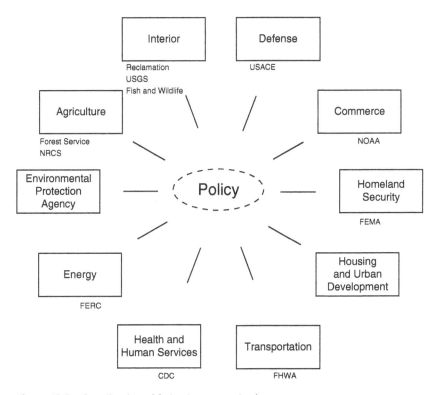

Figure 12.3 Coordination of federal agency roles in water management.

While there are many Federal water agencies, including some not shown on the diagram, those with major roles in water resources management will be discussed briefly.

The USACE and USBR are the main US federal water agencies that own and operate infrastructure, and both have long histories dating back to significant events in the nation's history and were organized to meet specific economic, social, and security needs of a growing nation. The USACE has projects and activities across the nation and in some other countries, while the USBR operates in the 17 western states of the United States.

Most divisions and districts of the USACE and the regions of the USBR follow river basin lines. Figure 12.4 illustrates the major river basins of the conterminous US, with some of the major USACE and USBR areas noted. The numbers of the river basins indicate regions assigned by the USGS. Full maps of each agency's operational areas can be obtained by Internet searches.

The USACE belongs to the Department of Defense, and it has a military side and a civilian side. The military orientation came about because navigation is a

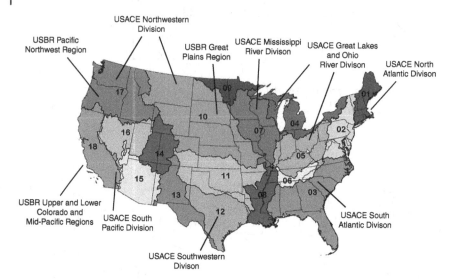

Figure 12.4 River basins of the United States with USACE and USBR areas shown.

national and homeland security issue. The civilian side is organized under a general officer who commands the agency, and it has a command structure of military officers ranging from a commander of the Directorate of Civil Works to commanders of its divisions and districts.

The Bureau of Reclamation is in the Department of Interior, and it owns and operates water reservoirs and facilities in the western states. It has a Washington DC office and technical offices located in Denver. The Bureau, as it is sometimes called, was organized to provide irrigation water and security for westward expansion in the nation. The Department of the Interior houses several other agencies that also addressed needs of the expanding nation, such as forestry and fish and wildlife. Currently, the same needs are felt throughout the country where the Department of the Interior operates other programs.

Over the years, there have been many discussions about whether the USACE should remain within the DOD and/or whether it should be combined with the USBR. While it seems inevitable that these questions will be raised again, the result always seems the same, which is to make no changes. An important reason for this is that both agencies have very strong constituent groups which will strongly resist changes in organization and authority.

The USGS has missions in water, geology, mapping, and other areas. The water program monitors and assesses water resources and is supported by congressional appropriations, matching funds from local, state, regional, and tribal agencies, and other work done in cooperation with various organizations, including other federal agencies, other countries, businesses, and academia. From time to time, politicians question whether the USGS is an appropriate federal activity and seek to cut its budget and mission. Their argument is that some of the functions, like mapmaking, could be handled by the private sector. However, water managers and other players in the water sector have a firm realization and recognition of the importance of sustaining USGS programs like stream gauging to serve everyone.

NOAA is housed in the Department of Commerce and hosts several water-related agencies, including the National Weather Service, which provides weather, hydrologic, and climate information and forecasts. This work occurs through national, regional centers, local, and river forecast centers. The NWS is discussed in the chapter about hydrology, flooding, and drought. The NWS could be located in a different cabinet department, but associating it with other earth resources agencies such as ocean research is logical.

The Environmental Protection Agency (USEPA) was created in 1970 and operates out of its Washington, DC, headquarters, regional offices, and laboratories. It has a regulatory mission, which for water issues is mainly conducted through delegated authority to state agencies. Prior to creation of USEPA, regulatory functions were spread among different agencies. The thinking was that regulation needed a higher profile and could work across different environmental areas symbiotically. The agency also conducts assessments and research, and it prepares useful educational materials. It employs many engineers and scientists, as well as other professionals in law, public affairs, finance, and information technology.

The Federal Emergency Management Agency (FEMA) is housed in the Department of Homeland Security, and was organized in 1979 to coordinate disaster response. The National Flood Insurance Program was already in operation under a predecessor agency when FEMA was created. By coupling flood insurance with disaster response responsibilities, the nation has a single agency focused on the action agenda for flood risk management. FEMA also response to other types of disasters. It gets most of its publicity from disaster response, but it also supports state and local governments with other services and training opportunities.

The Natural Resources Conservation Service (NRCS), originally known as the Soil Conservation Service (SCS), is in the US Department of Agriculture. The SCS was one of the New Deal agencies organized during the Great Depression, mainly to respond to soil erosion in the Dustbowl. Now, it mostly provides technical assistance to farmers and other landowners, with the general mission to conserve natural resources on private lands through cooperative partnerships with state and local agencies. Much of the private land is associated with agriculture, so although other agencies deal with natural resources, the NRCS fits well within USDA. The agency has also prepared engineering guidelines in the past and developed important hydrologic procedures. It conducts snow surveys and has funded a number of watershed improvement projects.

Many other federal entities have responsibilities related to water. The Department of Housing in Urban Development engages in housing and community development activities, which require safe and reliable water services. The Tennessee Valley Authority (TVA) is one of the most controversial federal agencies because it seems to compete directly with private power companies. During the 1930s, it created new opportunities and economic activity in a severely depressed area, but today its necessity as a federal entity is not as clear. The FERC is another example of a federal agency with substantial water responsibilities, but these are mixed with other energy regulatory functions such as electric power generated by fossil fuels.

For the most part, state water agencies were created to carry out drinking water and wastewater regulatory programs. Some of these, focused on public health, were implemented before federal activity ramped up to its present levels. Later, with the CWA and SDWA, greater authority was delegated to the state agencies. During the 1970s, many of them created water resources offices using funding from the Water Resources Council. Some of these water offices survived and, in some cases, were merged with water quality offices to create comprehensive organizations.

There has been a continuing debate about whether planning agencies should be combined with regulatory agencies because it seems that planning takes a backseat to regulation. In the western United States, the surviving water planning offices moved in some cases toward water development missions, and in other cases began to engage in modeling and surveys. Most western states have offices for water rights administration under state engineers or similar administrators. A few states have broader water responsibilities, especially in the West. The most prominent is the California Department of Water Resources, which administers a large program of water projects that were necessary to provide water to a rapidly growing state. The Texas Water Development Board is another example of a larger and more active state agency. The wide variation of climatic conditions and rapidly growing population have made more involvement in water management necessary in the state.

Basin and Regional Water Organizations and Authorities

Basin water organizations address issues that extend beyond political boundaries, like in river basins or other multi-jurisdictional regions. They can overcome the problem of mismatch between watersheds and political jurisdictions to some extent, but the jurisdictional authorities often resist cooperation.

River basin commissions that are for planning and coordination may lack much authority. In some cases, however, basin organizations can be authorities like regional multi-purpose water districts or utilities. An example of a planning commission is the Upper Colorado River Commission, which coordinates water allocations among four states in the Upper Colorado River Basin. The Trinity River Authority in Texas is an example of an authority that provides areawide water services to benefit other local authorities. The Rathbun Regional Water Association in Iowa is an example of a different type of areawide service organization as it serves about 80 000 people with water. Another example is the Mile High Flood District, which provides flood and stormwater services in the Denver region. There are many other examples of regional authorities, such as the California water districts. Also, the water management districts in Florida are special purpose regional authorities for flood protection, drought response, and delegated regulatory programs for water management.

Water Service Providers

Water service providers operate in spaces for water supply, wastewater management. Stormwater management, irrigation water, instream water, and flood protection. The most prominent are water, wastewater, and stormwater utilities, which own more infrastructure and serve more people than other types of water management organizations. The utilities fit the definition of a "public utility," which means an organization that delivers a "public service" for all members of a community.

While most water services are provided by government organizations, private water companies also provide some services. These can be totally private or investor-owned and are normally regulated by state public service commissions for rates and performance. Private water companies operate in most parts of the United States. An example is American Water, which owns and operates subsidiaries in different states. Sometimes parts of government organizations are also operated by private businesses, as in wastewater treatment plant operations, for example.

Water supply is the oldest of the water services and can serve as an archetype for the organization of wastewater and stormwater services. Public providers of water supply are usually departments of local governments, public authorities, or special districts. They are regulated only through political controls for rates. As monopolies, privately owned water supply utilities are regulated for rates by public commissions. All utilities are regulated for environmental and health purposes.

USEPA manages a classification system that was established by the SDWA for Community Water Systems and Non-Community Water Systems, which can be non-transient, like schools, or transient, like campgrounds. Water systems can also be classified by public or private ownership. Sometimes businesses will provide water to a local community as an ancillary service. Another variation is the consecutive service, where one utility provides wholesale water to another.

Organization of wastewater services is like water supply, but it lacks a similar classification system. USEPA does not have a mandate to count something like "community wastewater systems," but they do count treatment plants. Often the collection systems are separate and managed like drainage pipes in a city. Stormwater services are through local governments and sometimes through stormwater utilities. Their progress toward becoming "utilities" may have peaked.

In the case of irrigation, most systems involve diversion and distribution of water. In the West, the most common example is the mutual irrigation company, which is like a utility. These systems employ relatively few water managers as compared to public utilities. They were mostly created due to the "mutual" in the name such that local irrigators banded together to help each other.

Flood control is not considered a water service but is more like a set of coordinated activities among governments, individuals, and businesses. However, a few organizations are primarily for flood control, like the Miami Flood Control District in Ohio. Some multiple purpose districts bundle flood control with other water services. The water districts in Florida are examples of these.

Instream flows are not provided by single organizations but are handled by all water management organizations working together, and often are mandated by regulations. Organizations operating reservoirs have major responsibilities for instream flows as they coordinate the diverse needs and rights of water users and diverters.

Regulators

There should be a clear separation between regulators and water management organizations to avoid conflicts of interest. The main regulatory organizations are state agencies which oversee the SDWA and CWA requirements. Their authority is delegated from USEPA, which receives its authority from Congress. State government public service commissions regulate the rates and performance of private water companies. FERC is a regulatory agency in the sense of controlling permitting for private hydropower operations.

Support Organizations

The support sector of the water industry is important and active. Most support organizations are in the private sector, including consulting engineers, constructors, and equipment vendors. These businesses provide most of the planning, design, construction, and equipment for the water industry. Most jobs for engineers are in the consulting firms, who exert substantial influence on the water sector by planning, designing, and problem-solving.

Many nongovernmental organizations (NGOs) support the water industry, such as professional societies, trade associations, and interest groups. NGOs have major influence on water policy, such as the influence brought to bear by environmental interest groups, for example. Professional societies like ASCE also have major influence. AWWA is a combination of a professional society and a trade association. Its purpose is generally to support advancement of the water industry and this occurs by supporting members, like a professional society, and by providing a forum to exchange information about products and services.

Integration Among Water Organizations

Taken together, the fragmented organizations comprising the water industry struggle to take an integrated approach. Organizations naturally pursue their own objectives. This strong incentive drives the culture of the organizations and the individuals in them, who find advancement opportunities by pursuing single purposes. This creates stovepipes that are evident among organizations with different purposes, at the different levels of government, in different jurisdictions, and through different agendas of participants.

The fragmentation impedes progress in integrative areas such as IWRM, One Water, and collective action.

Examples of stovepipes in the water industry can be due to purposes (irrigation versus flood control), government level (national versus local government), jurisdiction (two cities competing for the same water source), and agendas of players (such as the farm industries versus environmentalists). The US Water Resources Council was a failed attempt to overcome fragmentation at the federal and state levels. Its story is told in Chapter 13 (planning).

Ultimately, describing the water industry as a collector rather than a sector helps to explain its fragmentation. The dilemma is, of course, what to do about it. Solutions are possible in local areas where different organizations and stakeholders find it in their individual and collective interest to work together. Case studies of successful efforts of this kind help to illuminate the range of possibilities for us.

Reference

Congressional Research Service (2017). *Selected Federal Water Activities: Agencies, Authorities, and Congressional Committees.* https://sgp.fas.org/crs/misc/R42653.pdf.

Questions

1   For a typical state government, identify a regulatory agency and a mission agency related to the water sector.

2   Name one of the main water associations. What is its purpose? Who are members? What is the benefit of belonging to it? What does it contribute to the water sector?

3   What does it mean that water is more of a connector among sectors than a sector itself?

4   What is meant by the term "water services?" Give three examples.

5   What is the Congressional Research Service and why would it write a report about Federal water agencies?

6   Does the US government have a water department? If not, how does it handle water management through other departments?

7   Explain how the USACE and USBR became the main US federal water agencies that own and operate infrastructure.

8   Is it logical that the USACE belongs in the Department of Defense? Why or why not?

9   What was the original main purpose to create the Bureau of Reclamation? Why is it housed in the Department of Interior?

10   What is distinctive about the USGS mission in water as compared to the USACE and the USBR?

11   Why would the National Weather Service be located in the Department of Commerce?

12   Is it logical that the Environmental Protection Agency is a separate agency, as opposed to having regulatory authority dispersed in other agencies?

13   What are the major functions of the Federal Emergency Management Agency that relate to water management?

14 What was the original impetus to create the Soil Conservation Service? What rationale would there be to change the name to the Natural Resources Conservation Service? If the Department of the Interior houses other natural resources agencies, why is NRCS in the US Department of Agriculture?

15 Explain why the original mission of most state water agencies was to focus on drinking water and wastewater regulatory programs.

16 How would you describe the long-term result of the Water Resources Council's programs to support water planning and state governments?

17 Why did California organize a large Department of Water Resources when other states did not do so?

18 What are the forces behind the need in Texas for a special agency named the Texas Water Development Board?

19 To what extent are basin water organizations able to overcome the problem of mismatch between watersheds and political jurisdictions?

20 What are some of the functions of the Trinity River Authority in Texas and what is your opinion about the need for them?

21 Did the USEPA staff conceptualize the classification system for Community Water Systems and Non-Community Water Systems or were they developed otherwise?

22 Give an example of a Non-community, Non-transient water system.

23 If they are about 50 000 Community Water Systems, is there a comparable number four wastewater systems? Why or why not?

24 Our stormwater services organize the same way as water supply services? Explain.

25 Why were mutual irrigation companies created and how do they work?

26 What is the main source of authority for state agencies which regulate SDWA and CWA requirements?

27 What is a state government public service commission, what is its authority, and what does it do?

28 Why would FERC be considered a water sector regulatory agency?

29 The greatest number of professional jobs in the water sector are among consulting engineers. How do they influence water resources management?

30 Why does fragmentation of water sector organizations block efforts to take integrated approaches to solution of problems and to foster collective action?

31 What is it about the incentives for water organizations and employees that blocks integrated approaches?

32 Give three examples of stovepipes in the water industry that comprise barriers to integration.

33 With fragmentation in the water sector, what is the best hope to forge more integration?

34 Should a water development agency be combined with a regulatory organization? Why or why not?

35 Why is the USACE still in charge of many dams already built? Cannot its functions be handled by the states?

36 Are research and data agencies like the US Geological Survey essential, or can their functions be handled by the private sector?

37 Sketch a simple organizational chart for a water supply utility and explain how each part helps it to achieve its core purpose.

38 What kind of organization was the US Water Resources Council?

39 Should a planning agency like the Alabama Water Resources Office be combined with the Department of Environmental Management?

40 What is the difference between a government agency and an NGO?

41 Is the work of TVA an appropriate government function?

42 Should river basin commissions have authority?

43 What is the principle of "subsidiarity?"

44 Explain how water services relate to economic "goods and services."

45 Explain what is the "service" provided in wastewater, stormwater, irrigation, and flood control.

46 Define a "public utility."

47 What are: CWS, NCWS, ancillary systems, consecutive systems.

48 Give an example of an investor-owned water company.

49 What is an NGO? Give an example from the water sector.

50 Give examples of: a special district, a federal mission agency, a basin organization.

13

Planning Principles, Tools, and Applications

Introduction

Water resources managers confront situations that require them to develop plans and strategies, both for needs of their own organizations and to work with others in collective action. These lead to a set of activities called "water resources planning." As used here, the term means planning to achieve objectives of organizations, planning with others for collective action, and planning to develop water infrastructure. Principles, methods, and tools for the first two activities are discussed here and planning for water infrastructure is discussed in the Chapter 14. The chapter after that focuses on how water resources planning can respond effectively to social needs, and planning scenarios are also discussed in other chapters, such as for conflict resolution and regulatory scenarios in the law chapter, planning for operational situations in the operations chapter, and financial planning in the finance chapter.

The Water Resources Planning Act of 1965 (WRPA) called for a comprehensive set of activities from national water assessments to planning local projects. While its programs are mostly gone, the activities remain important. They include assessing water resources needs; making framework plans for water management at national, regional, and state levels; preparing river basin plans; negotiating solutions to issues in shared basins; and planning solutions for local water issues.

Water resources planning is different from planning in other arenas because of the extent of collective action and public involvement required. People and communities must plan collectively because water resources provide shared benefits. Determining how to do this is a central purpose of water governance, which was discussed in Chapter 11. Water management is a public activity because most activities involve collective action, but it can also be private when decisions are about the water systems of private entities.

Water planning occurs in the United States within the federal system of government where the powers are distributed among federal, state, and local levels.

Water Resources Management: Principles, Methods, and Tools, First Edition. Neil Grigg.
© 2023 John Wiley & Sons, Inc. Published 2023 by John Wiley & Sons, Inc.

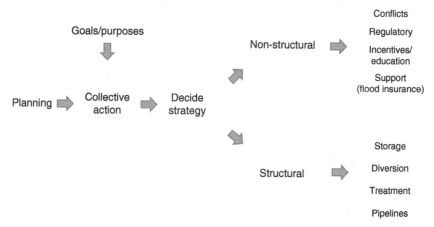

Figure 13.1 General process of water resources planning.

The federal government owns and operates many major water facilities, and it also regulates and drives policy. State government activities are mostly (but with some exceptions) regulatory, and local governments, where most infrastructure is located, are creatures of state governments with their own planning needs.

The general concept of the water resources planning process is shown in Figure 13.1. Goals and purposes drive the process, which is usually initiated when a problem or need arises. Planning involves collective action among stakeholders, who must collectively decide a strategy involving either nonstructural activities like conflict resolution or structural activities like developing infrastructure facilities. The collective action required can be within a jurisdiction, like a single city, or it can be multi-jurisdictional, like within a river basin.

While the nonstructural and structural activities are different, many planning principles and methods are similar, like the need for public involvement, for example. This broad use of principles and methods is evident in the WRPA programs listed earlier.

Chronology

The foundation for today's shared approaches to water planning dates from the 1780s with the US Constitution, which outlines the federal system of government. Over the years, the developing nation addressed roles of federal, state, and local governments in water management. Initially, the federal government managed major rivers for navigation, state governments had little activity until western states developed water rights systems, and local governments began to develop water supply systems to meet felt needs.

By the early 1900s, government activity in water resources management began to increase. The USACE was managing river control projects, the Bureau of Reclamation was organized in 1903, and the USGS program had been initiated. By the 1920s, multipurpose water legislation was addressing flood control and navigation, and the Great Depression then led to national planning for water resources. Western states had developed their water laws and many local governments had created water supply systems.

During World War II the focus shifted to emergency actions, but government activity picked up after the war to address built-up water needs. A commission chaired by former President Herbert Hoover delivered a policy report in 1950, and a Senate Select Committee on Water Resources produced a report which led to the 1965 Water Resources Planning Act and 1964 Water Resources Research Act.

The Water Resources Planning Act created a federal Water Resources Council and provided for comprehensive planning for intrastate and interstate water resources. The prescribed activities included a periodic national water resources assessment, planning studies, state grants to establish planning organizations, principles and standards for planning, and organization of river basin commissions. The planning studies were at Level A (framework and assessments), Level B (regional or river basin), and Level C (implementation) studies. The Council had an active program, but it was abolished during the Reagan Administration. The Water Resources Research Act was an experiment that included support for state water resources research institutes and a grant program. The program survives in a limited form through a diverse network of state water resources institutes and with coordination by the USGS.

As an environmental decade, the 1970s led to the CWA, SDWA, and additional environmental legislation. These are discussed in Chapter 20. After 1980, water policy had stabilized and there were fewer legislative initiatives other than amendments and shifts in implementation strategies. During the 1990s security and critical infrastructure received priority. After 2000, activities have focused on revisiting policies in arenas such as water quality, drinking water, multipurpose projects, dam safety, infrastructure renewal, and water finance. The result is that the United States has a comprehensive but piecemeal approach to water planning with many stovepipes and checks and balances.

Principles and Concepts

Water resources planning should integrate the activities of different entities over time, space, purpose, and level of work. This is necessary to implement collective action toward management of shared resources. Integrating over time refers to the stage of planning, like preliminary to final planning. Integrating over space considers coordination in river basins and among jurisdictions. Purposes involve plans to achieve goals such as in water supply or in multiple purposes scenarios. The level of work refers to governmental level and organizational level. The national, state, and local governments undertake different types of planning, and organizational planning extends from enterprise-wide activities to details of specific tasks like for operators, system engineers, or managers.

Due to its broad nature, water resources planning is practiced in different contexts, but it should be fit for the intended purposes, which determine its scope. Some plans address single system purposes such as water supply, water quality and wastewater, irrigation and drainage, groundwater, hydropower, navigation, flood risk reduction, or environmental water. Others address combinations of these to develop multipurpose plans. Some plans can be for specific management functions, such as policy, organization, implementation, and operations.

The purposes of plans lead to different terms being used to describe their processes and scenarios. Comprehensive plans and framework plans describe the scope of planning. Other plans relate to functions such as operating plans, development plans, and program plans. Still others relate to the stage of planning or time horizon such as preliminary plans and long range or strategic plans. Plans for areas or jurisdictions may be called watershed or river basin plans, or they could be regional plans. Many plans will include details about infrastructure, which become part of master plans.

The scale of planning determines how it is organized. Scale refers to the size of the problem area and aligns with jurisdictions of cities, counties, basins, states, and nations. It is useful to classify planning scenarios as in the Water Resources Planning Act programs: national level for assessment planning like a framework plan; state level for framework and strategy plans; basin level for integrated river basin planning; and project or program level, normally feasibility planning at the local level.

Conceptually, planning fits into the field of decision theory. Every planning problem involves a rational problem-solving procedure that seeks to find the best way to accomplish a goal. The basic steps involved are problem identification; goal setting; formulation of options; evaluation of options; decision making; and implementation. The rational approach works for structured problems at lower levels and for those involving technical issues without much controversy, although it sometimes seems that almost any water issue other than small scale technical designs will provoke some disagreement. In less-structured problems, aims are not always clear, and controversies and disagreements must often be resolved by considering the divergent agendas of stakeholders in organizing effective participation. An example could be a regulatory decision about a water quality standard with costs and benefits that are disputed by experts and might disrupt a local economy.

Participation in the planning process occurs as planning unfolds and stakeholders array themselves by interest groups and levels of influence (Figure 13.2). These

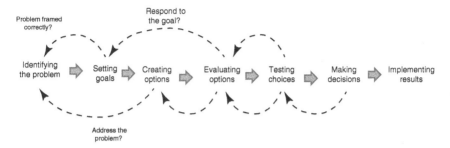

Figure 13.2 Feedbacks and iterations in water planning process.

positions vary over time, and stakeholders enter and leave the process. Intermediate and final decisions involve some or all stakeholders. These can involve meetings, reviews, studies, permits, and new developments, such as an election. Subprocesses such as identifying alternatives may be complex exercises in themselves. Outcomes produce alternative actions, which require negotiations.

The process begins with problem identification leading to goalsetting. In goalsetting, stakeholders may realize the problem looks differently and go back to redefine it. Once the problem is identified and goals are set, the team can define the system and begin to create options. Once they see what these options look like, they may question whether they really address the problem or if the problem is framed correctly. After a reasonable set of options is available, they can be evaluated, which requires collection of data about the options and inputs to simulation models to examine the behavior of the system under the different option categories. The simulation models and TBL assessment should begin to narrow the options to a favored few, which can be tested for feasibility and acceptability by the stakeholders. After the most feasible best choice is identified, the team is ready to recommend it leading to implementation. Feedbacks are experienced along the way, and it is also possible that the team will have to go back to basic problem identification and start all over again.

In some cases, a planning exercise may extend over a long time. In that case, interest groups and stakeholders may change from beginning to the end. If receiving permits along the way is involved, the government officials with authority to issue the permits may change and the planning group may lose some institutional memory. These factors add to the difficulty of planning in more complex situations.

Another aspect of planning is that it proceeds incrementally through stages, from the conceptual stage through renewal of the structure or program. These stages take on different names, depending on the scenario. This is particularly true for infrastructure planning, which has stages from the first reconnaissance to the final definite plans.

Accommodating the divergent agendas of stakeholders requires an effective approach to achieving consensus. This does not always require full agreement and different levels of consensus can be defined. For example, the best level is that a stakeholder can fully agree with the decision and accept it as an expression of the wisdom of the group. Another level is that the stakeholder can live with a decision but is not enthusiastic about it. This means that the stakeholder does not agree with the decision but will not block it. If the stakeholder's disagreement is stronger, this might result in blocking the decision and calling for more unity in the group and further work before consensus can be reached.

Scenarios

Many planning scenarios of water resources problems occur. As shown in Figure 13.3, the highest-level activities are policy and framework planning, and other scenarios involve planning for program development, infrastructure, basin coordination, and conflict resolution. Models and assessment tools are used extensively in the scenarios, which form archetypes or common patterns for situations the water manager faces.

Policy planning determines what a government or any organization intends to accomplish. A national government sets water policy, which establishes in broad

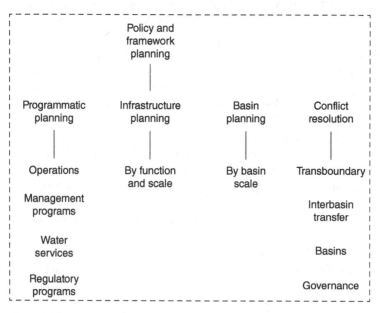

Figure 13.3 Levels and types of water resources planning.

terms how water should be used to advance national welfare and safeguarded. At a state level, the same concerns will be reflected in plans to address specific issues, like in a state in an arid zone versus one in a humid zone. At the local level, the board of directors of a watershed authority might set policy for how water can be used within its boundaries or for a strategy to raise funding to pay for its programs.

Policies are normally stated in water framework plans for nations and states. Large countries like the United States or a federation like the European Union will organize water policies at a high level without as many specific goals as in a state-level water plan. The United States is a large country, and state plans might resemble national plans for small countries.

In the United States, the WRPA envisioned a national water assessment, and two of these were completed. Many separate studies were conducted for sector and regional needs, but water issues in the country are so diverse that it is not feasible to assemble a national water plan for the United States. It is more feasible to have national water plans for countries that are more like US states in size. In these cases, a national water plan will normally include elements like policy for a multi-sectoral approach, a planning timeframe, goals for managing quantity, quality, and ecosystem needs, balancing supply and demand, water efficiency, and water security. Some countries have national water information systems to support their planning and management programs. Normally, these national or framework plans will identify the setting and issues to be faced, like drought and need for water supply; provide hydrologic information like precipitation and water resources; present infrastructure information like gaps and shortfalls; and identify policy needs, roles and responsibilities, and needed strategies. This type of information can also be found in some state water plans in the United States.

Water organizations undertake many programmatic actions that require plans, like rate-setting, incentivizing conservation, managing data systems, and operating flood warning and insurance programs. They also engage in budgeting, which is a form of planning. Capital budgeting is connected directly with infrastructure planning. In regulatory organizations, planning scenarios might focus on developing rules to implement policy. Operations planning creates plans for operating reservoirs, treatment plants, or other infrastructure systems. Examples include operating a reservoir, water or wastewater treatment plant, water distribution system, irrigation system, or well field.

Basin planning extends in scale from small watersheds to large river basins and regional planning includes areas such as metropolitan regions and specific urban areas. Planning for these provides mechanisms for shared governance to implement collective action. At different scales, stakeholders work within coordinating structures to address their diverse needs. The coordinating authority establishes basin boundaries and identifies stakeholders. Works and programs required are identified to set in motion planning programs for the future. Financing studies indicate payment shares, and partnerships and consortia are organized to undertake projects. Planning for estuary restoration is like a river basin plan because watershed actions determine estuary health. Planning for lake recovery is like estuary restoration but at a smaller scale.

The coordinating authority for basin planning, such as a federation of governments or an established water authority, should establish the basin boundaries and identify stakeholders. Then, it should organize a coordination mechanism like a river basin council with appropriate authority to create the plan.

Stakeholder needs for water should be compared with the supplies available. Then, works required (dams, etc.) are identified to set in motion planning programs for the future, including capital investment programming. The authority identifies environmental assets, impacts, and mitigation plans. Financing studies indicate payment shares, and partnerships and consortia are organized to undertake projects. The implementation phases include mechanisms to maintain the plan, inform and involve the public, and monitor and assess results.

Transboundary conflicts are like river basin and watershed planning, but the planning involves mediation. Inevitably, these encounter roadblocks that require conflict management. Parties may not see themselves as part of the same watershed due to different political jurisdictions or other factors that separate them.

Tools for Successful Planning

The planning scenarios outlined require use of tools to foster collective action, like public involvement, shared vision planning, and negotiation, among others. These can be used within different frameworks to approach planning problems, like a community-based problem-solving model developed by USEPA. Other tools are for analysis purposes, such as modeling or economic assessment.

When making plans, water managers can draw from a set of decision analysis methods and tools which seek the best ways to transform inputs into outputs to accomplish goals. The methods and tools fall into a few groups which align with the basic problem-solving process that is embedded in all planning scenarios. These groups of planning tools are used for framing problems, developing options, simulating what will happen, analyzing choices for decisions, and testing feasibility of the chosen options.

One of the challenges in developing plans is to frame the problem to be addressed and determine exactly what must be done. Problem identification in this stage is the first step in the planning process and planning should not proceed further until the problem is clearly stated and understood. As a decision support tool, the concept of problem identification is used in different fields which face similar problems. In the case of a mechanical system, problem identification would involve elements such as goals, customer expectations, and other performance variables. In the case of water resources problem, such elements might apply to an infrastructure component, but most problems involve more issues and problem identification is more complex, especially due to views of different stakeholders.

Sometimes problems are easy to identify, such as in structured water resources problems with fixed requirements, such as to meet a regulatory goal which may be nonnegotiable. However, many problems are unstructured and problem identification is more complicated. In these cases, an iterative approach is required. It will begin with a statement of the problem which may be accepted by some stakeholders but not by others. Collaboration and negotiation among stakeholders will then be necessary to examine whether the problem is correctly framed or not. For example, some stakeholders may feel that a problem requires immediate attention and others will think it can be deferred. In this case, the consequences of the current situation, the goals for solutions, the full range of stakeholders and responsibilities, and some idea of the strategies and options that might be pursued should be on the table for the planning group. The convening authority should work toward a clear statement of the problem before the next steps are taken.

After problem identification is completed, the next step is to develop options through a creative process to discover different ways to solve problems and achieve goals. There are different types of options in the technical, managerial, financial, and other categories. Normally in water resources planning, the technical options will come first. For example, if the problem is a polluted stream one technical option might be a treatment plant and another might be to relocate discharge points for wastewater effluents. Such technical options must then be evaluated with performance variables to determine their effectiveness, costs and benefits, and impacts.

After problem identification is completed, the systems of interest can be defined, and simulation can be performed using models and data analysis to determine how they behave and values of performance variables for different options. This requires formulation of scenarios to provide data for projections and models. Scenario development is used across many industries and situations, where it means considering alternative technological, social, and environmental futures to test what might happen under different plans.

Tools for analyzing choices begin at the general level with triple bottom line (TBL) approaches to consider economic, environmental, and social outcomes of the options to be studied. The TBL concept stems from the bottom line of financial reports but brings in criteria beyond financial results by including the economic, social, and environmental outcomes of courses of action.

Table 13.1 Sample layout of a TBL report.

	Economic impacts		Social impacts		Environmental impacts	
	(+)	(−)	(+)	(−)	(+)	(−)
Proposed action 1						
Group 1 impacts						
Group 2 impacts						
Group ... impacts						
Proposed action 2						
(add options ...)						
Proposed action *n*						

Table 13.1 shows a simple layout for a TBL report that shows positive and negative economic, social, and environmental impacts of proposed actions, with information included for different groups that will be impacted.

Multi-criteria decision analysis (MCDA) can be used to examine different weightings for the scores in the TBL report. In the TBL table, you can provide a net score for each project in each category. This requires scoring systems for assessing economic, social, and environmental impacts. In some situations, risks of different options may be reported as well.

Two methods are in general use for economic assessment to compare structural and nonstructural alternatives on the basis of commensurate costs and benefits, benefit–cost analysis (BCA) and cost effectiveness analysis. In BCA you examine benefits compared to costs, either as a benefit–cost ratio or net benefits. Cost effectiveness analysis is to determine the lowest cost solution to a fixed goal, like a regulatory requirement. These methods are explained in more detail in Chapter 18.

Environmental assessment is a broad category with the goal to determine net environmental impacts of options. These can be complex when all categories of natural systems are considered. In many cases, it is difficult to quantify the impacts. For example, if the flow in a stream is changed due to construction of a reservoir, analysis of the impact on different species of fish in their different life stages is required. In the case of US federal projects, an environmental impact statement or environmental impact assessment may be required. These are discussed in Chapter 20. If the project is not subject to federal laws, then an environmental or sustainability assessment may be required per local requirements.

Social impacts of options include categories such as health, income, and living conditions. These can be tricky to measure, and it can be difficult to achieve consensus about them among stakeholders. Social impacts and analysis procedures are discussed in Chapter 16.

Tools to test feasibility of options focus on involving stakeholders, especially public decision-makers, in consultations about whether decisions can be accepted. Feasibility is measured in the same categories as the options, like technical and financial feasibility. Review panels and individual discussions with experts and decision-makers will be useful to test feasibility. Public involvement is used to help achieve a method to foster consensus in decisions affecting multiple stakeholders. Methods of facilitation are useful in public involvement settings.

Questions

1 What happened to the federal Water Resources Council?

2 How does the Water Resources Research Act affect water planning and what is its status?

3 What is happening relating to national water policy post-2020?

4 How does the US federal system of government affect approaches to water resources planning?

5 How did roles of the federal, state, and local government evolve for water resources management?

6 What were major initiatives in water planning during the Great Depression?

7 Explain the role of the Senate Select Committee on Water Resources in creating the Water Resources Planning Act of 1964.

8 What do these terms mean in the context of water resources planning: approach, procedure, process, method, and tool?

9 What is included per the WRPA for Level A, Level B, and Level C plans?

10 Give examples of the scale, level, and stage of planning situations.

11 How does the planning for structural solutions differ from non-structural solutions?

12 What is meant by the concept of "problem of fit" and how does it affect planning?

13 In what sense is planning a "coordination mechanism?"

14 Give examples of "structured," "unstructured," and "wicked" problems. How do these relate to the concept of complexity?

15 How does planning relate to the concept of IWRM?

16 Explain the role of intergovernmental cooperation in water resources planning.

17 How does the concept of collective action relate to water resources planning?

18 What is the "Tragedy of Commons" and how does it relate to water resources planning?

19 What are the main provisions of the Water Resources Planning Act of 1965?

20 Identify the four levels of plans outlined in the Water Resources Planning Act.

21 The rational model is good for reasoning, but how does the actual political model of water resources planning work?

22 Give three examples of planning scenarios for collaborative problem-solving.

23 Give two examples of organizational planning in the water sector.

24 What is meant by scenario planning in the water sector?

25 Describe a situation where public involvement is important in water resources planning.

26 Give an example of risk analysis as it might be used in water planning.

27 How does the TBL concept relate to assessment of water resources plans?

28 Explain how an MCDA matrix can be used in TBL analysis.

29 What are the basic steps of the "rational model" of problem-solving?

30 Give an example of a coalition or interest group that might be active in water resources planning.

31 Outline how there are different levels to measure the degree of consensus among groups.

32 What is meant by shared governance and how does it relate to shared vision planning?

33 Explain the words collaboration and facilitation and give an example from water resources planning.

34 Give examples of indicators to report outcomes of feasibility studies in the financial, technological, environmental, and political categories.

14

Planning for Water Infrastructure

Introduction

Water resources managers and design professionals are often involved in planning for development of infrastructure water systems such as dams, treatment plants, and distribution networks. While these systems have different functions, the process to plan them has similar steps that follow accepted project development guidelines. This chapter explains the steps, purposes, and basic procedures of this planning process, which are mostly like those for other infrastructure categories, but involve special situations relating to the uniqueness of water projects.

The planning processes take on different forms depending on the types of water infrastructures and management organizations. Although some organizations have well-developed formal planning processes, others use informal approaches. Despite these differences, the process normally includes a logical sequence of stages from conceptual through definite planning.

The planning processes for small- and large-scale infrastructures are different. The small-scale example is represented by urban water systems, which are generally planned through the regular planning and budget processes of cities. For large-scale infrastructure, the process is exemplified in planning for multipurpose dams, which has drawn great scrutiny about economic-social-environmental outcomes, social issues, and safety.

Planning for water infrastructure also differs according to type and scale of system. Dams normally impact entire watersheds and geographical regions, while urban water and irrigation systems have smaller footprints. Some infrastructure systems focus on the site level for individual facilities and buildings. Small-scale systems and components of larger systems may be planned in conjunction with budget processes, while large-scale dams and larger urban water systems require more formal processes.

Water Resources Management: Principles, Methods, and Tools, First Edition. Neil Grigg.
© 2023 John Wiley & Sons, Inc. Published 2023 by John Wiley & Sons, Inc.

The process to plan water infrastructure components or systems differs from those to develop nonstructural programs. Decisions to construct infrastructure projects have longer-lasting consequences and obligations as well, so planning processes often attract more scrutiny than those focused on programs and even on basin planning.

Water infrastructure systems are capital-intensive and long-lasting, so they may be difficult to finance and implement. Moreover, they may be controversial and require extended processes of public involvement. Planning of water infrastructure projects is different from planning for non-structural programs and solutions, which focuses more on the establishment of management programs and the resolution of conflicts.

Planning across the Infrastructure Lifecycle

The process for planning water infrastructure considers sustainability through its life cycle, or the range of activities from initial planning through operations and eventually to replacement or decommissioning. As Figure 14.1 shows, planning during the initial phases focuses on new or expanded infrastructure, whereas

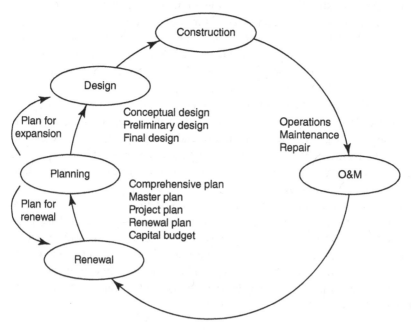

Figure 14.1 Infrastructure life cycle.

renewal requires a different approach to planning that focuses on decisions about whether to replace or rehabilitate facilities. The infrastructure life cycle includes phases with nuanced activities such as different stage of planning and design.

Infrastructure Planning Process

Planning practices differ by organizational types and sizes, such as among city departments, authorities with independent boards, privately owned utilities, national entities like USACE, and organizations with global perspectives, such as the World Bank. Embedded in all such planning practices are the steps of the general problem-solving process: problem identification, goal-setting, formulation of options; evaluation for feasibility, approvals and permits, finance, public involvement, and implementation.

Infrastructure project plans should fit into the hierarchy of comprehensive plans to master plans, capital improvement programs, and capital budgets. Project planning involves stages from conceptual to preliminary and then to final plans. These stages can also be named reconnaissance, feasibility, and definite project planning. In the conceptual or reconnaissance stage the need for a project is identified and the first set of options is developed. The feasibility stage explores the financial, technological, environmental, and political viability of the project. The definite project phase includes detailed examination of alternatives and their feasibility. This stage is also used to develop construction plans, specifications, and operating agreements. Implementation as the final stage involves construction and startup.

Project Development Guidelines

Established organizations have project development guidelines, which are often presented in formal manuals that contain policies and procedures. These range from large and sophisticated organizations to smaller utilities and local governments and will align with the purposes of the organizations. For example, at the large investment bank level, like the World Bank or Asian Development Bank, project development guidelines will take on a regulatory stance. The investment bank will assure that project funds are spent according to agreements and that integrity and quality are assured throughout the development process.

In the case of a project using funds from a development bank, the executing agency building a project will follow its own project development guidelines. An example of this approach could be the national water authority of a small nation in Africa or Asia that is building a dam with World Bank support. That authority will normally have other projects in the development stage, and all should comply with its own policies and procedures. Of course, these must be consistent with World Bank requirements for the project involving its funds.

Established organizations like the USACE and the Bureau of Reclamation have formal procedures that have been developed over many years and through experience. The USACE has a "Planning Community of Practice" with planners representing different parts of the organization. Participants serve on multi-disciplinary teams to focus on plan formulation, economics, ecosystem restoration, and cultural resources. Development of project guidelines will be advised by the planning community, and policies and procedures will be maintained by the organization.

The USACE process follows six steps that encompass the conceptual to feasibility stages: inventory and forecast; plan formulation (objectives and constraints, measures, and alternatives); plan evaluation; plan comparison; and plan selection. To streamline the process, the USACE adopted a process named "smart planning," which was intended to reduce unneeded complexity and delay in projects that involve multiple parties, complex regulatory structures, and possibly needing conflict resolution. It requires a maximum of $3 million spent over 3 years with 3 levels of enhanced vertical teaming ($3 \times 3 \times 3$ rule). Length and format of feasibility reports are also specified. The USBR has also developed extensive guidelines and has a manual with many parts that can provide useful insights about project development and management.

At the level of utilities and smaller organizations, development of project guidelines is a cottage industry where each organization develops its own procedures, although sometimes it may adopt the basic guidelines from another organization. When a water organization develops its own guidelines, it can find extensive resources by studying documents available in well-established agencies. Many of these can be found through an Internet search using terms like "project development manual," or "project management manual."

Planning Phases from Conceptual to Final Plans

The rationale for the conceptual or reconnaissance stage is that the need for a project is evident, but different ways to meet it should be explored. The scope of the plan at this stage will depend on whether the work will focus on screening multiple alternatives or whether a definite approach has been selected and conceptual plans and information are needed to facilitate review and decision making.

By short term and approximate studies of these alternative ways to meet the need, the executing organization can screen them to determine the best general way to proceed. The studies should be in enough detail, of course, to highlight the major strengths and weaknesses of the alternatives. The information to be included in the plan should comprise objectives, ways that they can be met, the technical and programmatic approaches, costs, risks, barriers, procedures, stakeholders, and other relevant information to help determine whether to proceed to the next level. If a definite pathway has been selected, then the same information should be included but with more information.

Feasibility studies follow naturally from the conceptual stage as the responsible organization explores whether and how to undertake project. These should consider economic, technical, legal, scheduling, and other factors to determine whether a project can succeed. As an example, a local government might follow up on a conceptual watershed study to determine feasibility of acquiring property to create a multi-use facility.

Elements of feasibility studies should explore whether the public wants the project and the associated political aspects. For example, are there properties needed for acquisition where owners will resist selling and create unforeseen obstacles? Affordability and financial feasibility are important because they can make a project succeed or fail. Legal feasibility will explore issues and options to deal with regulatory issues and other matters of law, including the possibility of litigation. Environmental feasibility is closely linked to political feasibility as the public will be aware of negative impacts. Technical feasibility is, of course, essential because the project must be constructed, and the project must ultimately achieve its purposes.

Final or definite project plans (sometimes stemming from what is called the design phase) include the detail needed to support and construct the project and ensure that it can be operated to meet its basic purposes. These details can include extended studies and reports about each aspect of the project, including technical design, financial arrangements, property acquisition, operational guidance, quality assurance, and review. Each of these categories will have been explored during the preliminary phase, and the final planning or design phase establishes the definite procedures and decisions involved. For example, the size, configuration, and operating schemes for the infrastructure system will be established. The financial details will be made definite, such as source of capital and operating funds. Operational guidance will be coordinated with the technical design to ensure that project operation can be successful.

Example: Large Scale Dams and River Works

Large-scale dams and river works are developed by various national and state governments, as well as by special authorities. The Corps of Engineers and the Bureau of Reclamation are the main dam construction agencies in the United States, whereas in other countries national water authorities are often involved.

Dam construction in developing regions is normally led by these government organizations, often in cooperation with donors and private power operators.

The traditional process for planning of dams was developed during the early part of the twentieth century. As a robust developing nation, the United States planned and built many dams. Other countries came along later, but a common process was established to include steps for hydrologic appraisal, reconnaissance, detailed investigation, definite planning, economic planning, financing and implementation studies, and impact assessment.

Subsequently, environmental and social concerns about dams have raised the bar for the planning process and it has become difficult to obtain approval for them in the United States. However, as explained in Chapter 3, a number of dams are being developed in emerging countries. A World Commission on Dams was convened and reported in 2000 about the needed attention to social and environmental concerns. The Commission developed a set of principles for planning of dams that included attention to these concerns, as well as technical issues. Details would include a focus on benefits and beneficiaries on income, gender, and ethnicity groups. Also, the plan should align with policy at national or regional levels and with river basin plans. In the event that a dam affected indigenous people, plans should confirm that they were consulted and their views considered in a meaningful way. If the dam would trigger transboundary issues, permission from stakeholder countries would be required. A number of other technical and safety issues must be addressed as well.

In the United States, it is apparent that few new large-scale dams will be constructed. Moreover, there is controversy about whether some of the older ones should be removed. Dam removal is an active topic of discussion in the United States, particularly where safety and environmental issues are paramount.

Urban Water Systems

Planning of the infrastructure for urban water systems is a common scenario because so many cities and communities require expanded or renewed drinking water, wastewater, and stormwater facilities. These may be managed by general governments or separate utilities. In either case, they must be planned through the ongoing urban planning systems of the community. This is an example of the hierarchy of planning where project plans should fit within comprehensive urban plans and appropriate master plans.

As the community develops, expenditures for all infrastructure are programmed through a capital improvement program process and then funded through capital budgets. Once a project is budgeted through a capital budget, the funding will begin to flow, and procurement can be initiated. This will include acquisition of design services, equipment, and construction. The procurement of planning services is normally based on a request for proposal or RFP. These planning services may involve successive phases through feasibility analysis to final design.

A somewhat standard protocol and sequence for developing urban water infrastructure is practiced in the United States. While details may differ, the following example scenario occurs repetitively. A community is growing and lacks sufficient water supply. The community's decision authorities decide to act and they direct its managers to either prepare a plan or retain a consulting firm to prepare one. Depending on the complexity of the project and the community procurement rules, the managers either retain a consultant directly or develop a request for proposal.

The project will require a plan of work and, once a consultant is retained, the community will execute a contract to specify the scope of work, schedule, and compensation. The consultant will evaluate options, including new supply, new infrastructure, conservation, purchasing water from another city, trying to get the federal government to build a reservoir, and others. After considering all options, the consultant will make a presentation to city decision-makers and be directed to explore a selected set of options in detail.

After further study, the consultant will return with the next phase of plans. After a decision process, the city will choose a plan and direct the consultant to prepare conceptual or preliminary plans. Depending on the level of detail, these may be adequate to apply for permits, or additional studies and submittals may be required. At the end of the project development phase, construction plans will be prepared, and construction will be authorized.

Questions

1   What is unique about planning for water infrastructure projects compared to other physical infrastructures?

2   What is meant by a system being capital intensive?

3   What are examples of comprehensive, multi-sector plans where water infrastructure projects should show alignment at the national, state, and local levels?

4   Conceptualize a type of water infrastructure project and give examples of the following steps in the planning process: problem identification and goal-setting; formulation of options for infrastructure types and alternative configurations; evaluation for feasibility on a multi-objective basis; approvals and permits; finance; public involvement; and implementation.

5 What is meant by a reconnaissance stage for an infrastructure project?

6 Give examples of indicators to report outcomes of feasibility studies in the financial, technological, environmental, and political categories.

7 Give explanations of what is involved in design plans, specifications, and operating agreements.

8 Explain the construction process to include bidding, review, award, organization, construction, inspection, and acceptance.

9 Explain the difference in planning for a dam during the older "big dams" era and today.

10 Explain the social issues that may arise in planning for a contentious infrastructure project. Use a hypothetical example of a wastewater treatment plant.

15

Water Quality Planning and Management

Introduction

Water quality in the environment depends on multiple human and natural effects, and it requires a comprehensive picture to show how the pieces fit together so an integrated approach to managing it can be mounted. Drinking water quality also depends on controls within treatment and distribution systems, which are discussed in Chapter 20. This chapter focuses on water quality in the environment.

The goals of environmental water quality management face many conflicting demands, lack of consensus about what comprises good water condition, complexity in assessing the condition of water, and in many cases by poor governance in setting and maintaining water standards. These challenges are formidable because surface waters are used and reused for many purposes as they pass through the hydrologic cycle and groundwaters are subject to abuse unless managed carefully. Nonpoint sources pose an even greater challenge. Water quality planning can provide a coordination mechanism to orchestrate cooperative efforts to achieve the needed collective action.

Considering these issues, the organizing questions for this chapter are what is the status of environmental water quality, how can it be improved, what makes an effective water quality management process, and what are the roles and responsibilities? The chapter is a companion to Chapter 8, which explained how water quality is determined and how it links to human and ecosystems health, as well as Chapter 20, which explains the laws that regulate water quality. These other chapters show how both natural and human impacts cause changes that affect water quality and how society's responses to these changes have occurred through law. It extracts from these and other chapters to offer a roadmap for how water quality improvements can be planned and managed in today's institutional environment. This institutional environment is based on the Clean Water Act to address wastewater discharges and environmental water quality, the SDWA to provide tools for

Water Resources Management: Principles, Methods, and Tools, First Edition. Neil Grigg.
© 2023 John Wiley & Sons, Inc. Published 2023 by John Wiley & Sons, Inc.

the quality of drinking water, and on additional laws which require comprehensive planning, such as the Federal Power Act.

A starting place for the discussion is to identify what is meant by water quality problems. This might seem like an obvious question, but it is not obvious because perception of water quality is to some extent in the eyes of the beholder. Water quality is difficult to measure due to its many attributes, but generally think of it as an unacceptable water-related situation that causes harm and requires a response and the measurement would be related to the seriousness of the problem. For example, if a heavy rain washes sediment into a river but it passes soon, it produces a change in water quality but there is little that can be done about it in many situations. On the other hand, if a levee holding back a waste pond breaches, contaminates the water, and makes people sick, this is an unacceptable situation that might be avoided.

Status of Water Quality

Prior to the 1972 Clean Water Act, the United States faced the dilemma of finding a comprehensive approach to water quality management. Wastewater treatment was sparse and ineffective, nonpoint source controls had hardly been implemented resulting in deteriorated streams often filled with trash. There were only weak incentives and there were few legal or financial resources to approach the problem. A comprehensive approach was needed to address these issues with a feasible program.

After the law was passed there was a flurry of activity to determine how to implement its provisions. This was encouraged by availability of federal funding for states and communities, as well as for engineers, contractors, and vendors. By the late 1980s, amendments had focused on some of the issues, but it seemed that a more holistic and effective approach was needed. This led to a study named Water Quality 2000, which was conducted by public, private, and nonprofit groups and facilitated by the Water Environment Federation. Their 1992 report, "A National Water Agenda for the twenty-first century," identified causes of water quality problems and pathways to solutions based on pollution prevention, individual and collective responsibility for water resources, and watershed planning and management.

The recommendations were for logical but challenging actions, such as public commitment through education and training, preventing pollution from all sources, wise use of resources, managing growth and development, research and development, bridging policy gaps, incentives for clean water, and funding of programs. The report elevated attention to nonpoint sources, especially from agriculture.

Since the report, USEPA and USGS have issued periodic reports about the overall status of the nation's water. However, just as it is challenging to measure water quality, it is also difficult to provide a single picture of its status. For this reason, water quality reports tend to focus on individual watersheds and problems.

Science and Management of Water Quality

The science of water quality refers to how physical, chemical, and biological changes occur, and these are driven by societal forces like population growth and economic activity. A DPSIR framework provides the visual tool to illustrate the status (Figure 15.1).

Causes of water quality degradation are, like water quality 2000 reported, wastewater discharges, urban runoff, agriculture, roads and other transportation modes, and atmospheric deposition, including acid rain and nutrients. The drivers that increase these are global development, climate change, and rising aspirations. These changes are evident in point and nonpoint discharges from wastewater, industry, land use changes, agriculture, mining, energy generation, and other activities. The impacts of water quality degradation are hard on poor and vulnerable people, on fish and wildlife, on drinking water supplies, and on our access to healthy water bodies for various uses. Estuary degradation provides a clear picture, and is closely associated with problems such as the dead zone in the Gulf of Mexico. Another impact is deteriorated drinking water.

A great deal is known about the science of water quality. A starting point is that its elements are physical, chemical, and biological. Examples are sediment, dissolved

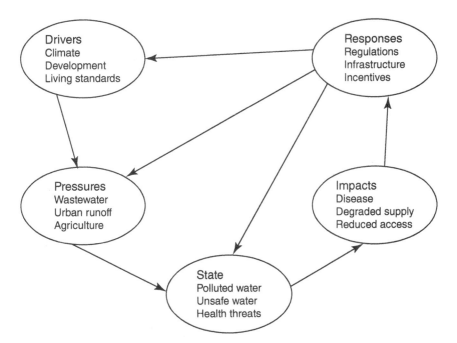

Figure 15.1 DPSIR depiction of water quality changes and responses.

oxygen, and bacteria. Details are explained in Chapter 8. We know that changes in the levels of these elements are due to human and natural causes. The natural causes are connected to hydrology and to biogeochemical cycling. The human causes are related to infrastructure and to our everyday activities or how we live.

In terms of engineering, the greatest advances in knowledge have been about transformation functions, or the effects of human activities causing pollutions and the systems of water treatment and mitigation that have positive effects on water quality.

Technologies for mitigation and treatment continue to advance. Analysis and assessment tools have also improved, and data, indicators, and reporting have provided much information to aid in management and regulatory decisions.

A great deal has been learned about health impacts of water quality degradation, as well as flooding. These are explained in Chapter 8, which presents six pathways whereby water management decisions and results impact water quality. The greatest impacts stem from degraded quality of drinking water, and these are especially hard on young children.

Water quality data can be divided into drinking water and environmental categories. Drinking water quality is regulated, and explained in Chapter 20. The way that the USEPA reports trends in environmental water quality is the identification of "impaired streams," which is provided by states by regulation. The top sources of environmental water impairment usually listed by USEPA by frequency of occurrence include sediments, pathogens, metals, nutrients, and dissolved oxygen. These lead to other indications of water impairment like fish consumption advisories and growth of noxious aquatic plants. The occurrence of sediment, pathogens, and nutrients suggests that agriculture is a big contributor, and other nonpoint and point sources from cities and industries add to stream pollution loads.

Management Functions

The CWA was a bold step forward toward improving water quality and might be considered as a first step in developing a comprehensive approach to improving it. It has become apparent that a comprehensive approach is needed and to accept that that it must be applied where water is managed. A way to view this is by identifying water "where it is," which was a basis for the concept of "Total Water Solutions" that was promoted by AWWA. For purpose of discussion, these locations of water can be used to visualize different management situations.

These situations are determined by the way that water moves from its natural sources through infrastructure systems and back to the environment. Figure 15.2, which was shown in Chapter 12 to illustrate the roles of water management organizations, can be adapted to show locations of influences on water quality.

Table 15.1 summarizes "where water is" (cohorts of water), management functions, and the control instruments. It provides a way to map how water quality

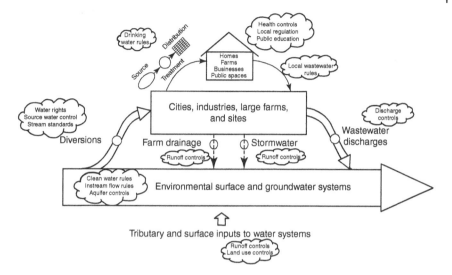

Figure 15.2 Water management influences on water quality.

Table 15.1 Management and control functions in water systems.

Cohorts	Management functions	Control instruments
Surface water	Instream flow management, reservoir operations	Classification of streams, water quality standards/monitoring, source water control, instream flow rules
Groundwater	Well pumping, recharge, aquifer protection	Regulation of well pumping, classification, permitting
Diverted water	Collection and transmission of water	Water rights and allocation controls Drought rules
Drinking water distribution	Store, transmit, treat, and distribute drinking water	Safe drinking water rules, multiple barriers
Household water supply	Operate/maintain building plumbing systems	Plumbing codes for drinking water and wastewater
Household sewage	Discharge of water to local sewers	Local rules about pretreatment and allowable disposal practices
Wastewater systems	Collect, treat, and dispose of wastewater	Wastewater discharge permits and controls
Runoff water	Operate stormwater and other drainage systems	Land use and management rules for nonpoint sources

management tools can be applied to control water quality. In turn, the mapped linkages point to responsibilities of different water governance authorities in management and regulation.

Comprehensive Water Quality Management Program

Managing the Quality of Water Where It Is

Policy designs under the Clean Water Act and its amendments provide the regulatory structure for water quality management, which must be translated into action through operating agencies to create a comprehensive management program that will address the quality of water "where it is." The big picture is illustrated by Figure 15.3, which shows the transformation functions of human activities and management controls that determine the condition of water bodies. Natural causes of water impairment are also shown, and while they can be addressed to some extent, the focus is on human causes.

Human activities comprise point and nonpoint sources of pollution. Point sources are discharged from the wastewater treatment plant or outfall pipe. Nonpoint sources include urban stormwater, septic tanks, roads, construction, farming, and other land disturbing activities such as forestry and mining. The Clean Water Act and other laws address each of these through different program elements and is the basis for a comprehensive program of water quality management.

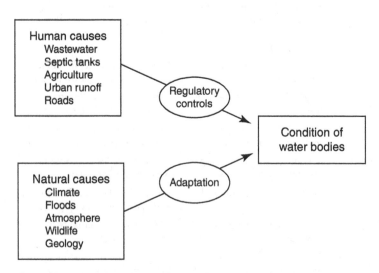

Figure 15.3 Human and natural determinants of water quality.

The main legal controls of the Clean Water Act were enacted in 1972 and have been modified several times through amendments to the law. As explained in Chapter 20, the 1972 Act was an amendment to the Water Pollution Control Act of 1948, the first major US law to address water pollution. The 1972 Act established the framework to regulate pollutant discharges, gave USEPA the authority to implement regulatory programs, established controls for water quality standards, required permits, subsidized construction of wastewater treatment plants, and emphasized planning for nonpoint source pollution. Subsequent amendments replaced the construction grants program with the Clean Water State Revolving Fund. Other laws changed parts of the Act, like in the Great Lakes Water Quality Agreement of 1978 to regulate toxic pollutants.

An overview of how the Act works is shown in Figure 15.4. Water quality management involves different players and actors, primarily those causing discharge of pollutants, those responsible to manage them, and regulators. The control system involves all three levels of government. Water quality standards are established by the states to determine required controls. States can assign different levels of standards, such as for drinking water sources, cold water fisheries, and others. Pollution reduction strategies for point source contaminants are based on the NPDES permit program, as explained in Chapter 20.

Nonpoint source controls involve different policy instruments. States receive grants under section 319 of the Act to manage their programs. The most direct control mechanism is the requirement for NPDES permits for stormwater discharges. This is a command-and-control policy instrument. Another instrument is provided by economic incentives which can be developed by different states. For example, developers could be provided with compensatory credits for pollution reduction activities. Education and volunteerism are important, especially to avoid degradation of water quality from land developments.

States are required to perform monitoring and assessment of water bodies and to submit reports. They must report to USEPA on the condition of the waters and

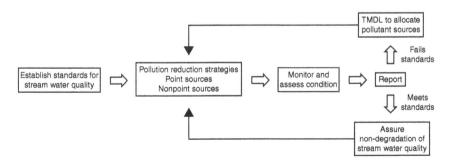

Figure 15.4 How the Water Quality Act works.

whether they are meeting standards on the basis of ambient monitoring. If they meet standards, only occasional monitoring is needed, and a rule called antidegradation is applied to prevent deterioration in stream water quality. If they do not meet standards, a program to establish total maximum daily loads (TMDLs) is required to guide allocation of the rights to pollute and development of a revised strategy. A section 401 state water quality certification program requires states to certify that water quality will not be impaired when permits under section 403 are issued. Also, a section 404 program regulates discharge of dredged or fill materials to wetlands and other waters. This program often leads to requirements for environmental assessments. The state revolving loan fund under section 603 provides access to financial resources.

Water Quality Assessment

The assessment of whether a state's streams are meeting the standards is based on the efforts by the states using monitoring data and any other methods they consider useful, including use of biological indicators. Under the CWA, states identify the water quality status of their waters and create a list of those that are impaired and require a TMDL evaluation. Another section of the CWA requires reporting of water quality in publicly owned lakes.

This comprises a specific application of the general concept of water resources assessment with a broader goal. On an international basis, the 1992 United Nations Conference on Environment and Development reported on the broad need for water resources assessment, including sources of supply, quality of water resources, and human activities that affect the resources. This holistic definition was to encourage nations to develop data management services to support water planning and management for socio-economic and environmental purposes.

This broader concept of water resources assessment provides a framework where water quality can be considered together with water quantity and any threats or changes that should be confronted. Terms used for condition assessment of water lack standardized definitions, but the general framework shown in Figure 8.5 illustrates the approach. Assessment is a process, it requires criteria, and the goal is to measure condition relative to the purposes of water use. This conforms to the general organization of the stream standards and assessment requirements of the Clean Water Act.

The general approach to determine condition of water relative to its intended uses considers both availability and quality of water. The adequacy of the water quality and quantity to meet needs, including for environmental systems, must also be considered.

National water assessment was required under the 1965 Water Resources Planning Act, which required the Water Resources Council to "… prepare an assessment … of the adequacy of supplies of water necessary to meet the water requirements in each water resource region in the United States and the national interest therein." Assessments were prepared for 1968 and 1973 under this Act, but the program ended and the 1978 Assessment was never published. Although this means that the United States lacks a formal national water assessment program, it has a piecemeal approach that starts with long-standing programs to gage streams by the USGS. The USACE also developed water assessments to support its river basin plans and policy analysis and the USBR also began basin assessments to support its development projects. After the demise of the Water Resources Council, USGS published a National Water Summary in 1984. This publication series was not continued, but USGS continues to evolve its reports for water resources assessment in separate studies and programs. Its legislation provides USEPA with assessment authorities across the environmental media and its reports shed light on the overall condition of the nation's water resources, despite not integrating state reports to create a composite picture.

Because the assessment of water condition requires standards for comparison, multiple criteria for different uses of water are required. This makes any universal assembly of criteria a hopeless task and has prevented establishment of a consensus measure for water quality. In addition, measures of water quantity and timing are required to assess suitability for natural systems, including timing of flows.

Technologies to Enhance Water Quality

As compared to 1972 when the Clean Water Act was passed, technologies are much better now. Instrumentation offers new ways to detect known and emerging contaminants with greater accuracy and precision, as well as to couple measurements with smart monitoring and control systems. Enhanced data systems coupled with improved models enable us to assess performance of systems and water conditions better. More research has been done on unit processes for point and nonpoint sources.

Some water quality measurements can be made directly in situ. Examples include temperature, pH, dissolved oxygen, conductivity, and turbidity. Instruments for these are being improved continually. Other water quality measurements require grab samples, which can be collected using different means. These base statistical challenges to be representative of the true state of water quality, so some instrumentation is available for automatic sampling or to impose other statistical controls on the data. Laboratory instruments are also improving and developing new capabilities to measure heretofore unrecognized contaminants.

Water quality modeling has improved and, together with better data, is more reliable than it was decades ago. This is particularly important for regulatory controls, especially the TMDL program, which will determine control strategies. Water quality databases have improved, including through volunteer efforts to develop a national stormwater quality database.

New and improved unit processes for water and wastewater treatment are being developed, and best management practices for stormwater control have been tested. There is an International Stormwater Best Management Practices database which can guide communities about their choices.

Lessons Learned from Experience with the CWA

Understanding how a comprehensive water quality program should work requires knowledge about the CWA, and USEPA provides educational materials to explain how it works through its website for the "Watershed Academy." Because water quality, is difficult to assess, no consensus national assessment of the effectiveness of the Act have been produced. However, studies have sought to provide partial views and there have been, of course, reports by advocacy groups about environmental benefits and costs of compliance.

Data will show that the Act has been successful at improving the condition of waterways based on individual parameters such as sediment and dissolved oxygen, but after decades of experience with the Act, the complexity and difficulty of assessing the overall condition of waters are clear. With ongoing development and climate change this problem will only grow in magnitude. Also, the economic and social benefits of water quality regulation will remain somewhat uncertain, thus giving fuel to opponents of regulation.

Questions

1 Identify the main point and nonpoint sources of water pollution. List the five types of contaminants that are reported most frequently in state reports to USEPA.

2 What were the pressures on water quality increasing and if so, what are the driving forces?

3 Identify the main negative impacts of water quality deterioration.

4 What was the status of wastewater treatment prior to the CWA?

5 Explain how you would identify what is meant by a "water quality problem."

6 Why is it difficult to define and measure water quality?

7 Characterize the status of the US national strategy for water quality management as it existed prior to the 1972 Clean Water Act.

8 What was meant by the WQ 2000 vision statement of "Society living in harmony with healthy natural systems"? How does this relate to the concept of sustainable development?

9 WQ 2000 concluded that fundamental changes in institutions, business, government and individual lifestyles will be required to enhance water quality. Do you agree, and what would be some of the changes needed?

10 Discuss the advantages and disadvantages of a regulatory versus market-based approach to water quality management.

11 Under the CWA, who has responsibility to set water quality standards? Should they all be the same or what would cause them to vary?

12 Do you consider that the Clean Water Act is designed well? If you were to redesign it, what would your changes be?

13 Explain how the NPDES program works and relates to monitoring of water resources.

14 How does the 404 program of the CWA affect the siting of new reservoirs?

15 Name the main policy instruments in place to control nonpoint sources.

16 Under the CWA, what are the provisions for controlling stormwater quality?

17 What is the purpose of the section 401 water quality certification program under the Clean Water Act?

18 Who assesses water quality in the United States, USEPA or state governments? What does USEPA require of the states?

19 Are biological indicators suitable to assess water quality?

20 If water quality in a stream in a state does not meet the standard, what is the required remedial action?

21 What is your philosophy on the need to subsidize wastewater treatment? Do you believe the funds spent by the United States on the Construction Grants Program were well-spent?

22 What is meant by "multiple barriers" to protect drinking water quality?

16

Planning for Sociopolitical Goals

Introduction

Among the many facets of water management, the connections of people with water stand out as most important. This importance is clear when you consider our dependence on safe drinking water before any other human need. The connections go on, of course, to the need for food, flood security, hygiene, energy, and many other needs that depend on water. These strong links cause people to have emotional connections to water, which will be manifest in water resources management situations.

These emotional connections mean that water managers will encounter many sociopolitical issues as people express strong feelings about water in political arenas. These sociopolitical issues can be addressed to some extent by making water plans equitable, including plans to sustain ecosystem services. Even when sociopolitical issues are not under the control of water managers, they can contribute by taking actions within their own sets of authorities.

Water management scenarios in every country include situations where injustice and inequities can lower the quality of life for segments of the population. People lose confidence in water utilities due to failures, shutoffs, boil water notices, red water or anything that makes their water seem bad. Other problems are unaffordable services, exposure to pollution, ineffective drainage systems, lack of irrigation water for smallholders, and flood risk. When confronting problems like these, water managers should create plans that consider public opinion and interest group agendas.

Sociopolitical goals involve broad issues like health, environmental problems, and related problems of central concern to people and politicians. Addressing these through water management goes in parallel with society's other efforts to address them through movements like sustainability, corporate social responsibility, and environmental, social, and governance reporting.

Water Resources Management: Principles, Methods, and Tools, First Edition. Neil Grigg.
© 2023 John Wiley & Sons, Inc. Published 2023 by John Wiley & Sons, Inc.

Water managers have been criticized for not being sensitive enough to sociopolitical concerns. To respond, they should collaborate with others who are experienced in sociopolitical decision-making, especially learning from social science experts. This chapter summarizes available knowledge about sociopolitical water issues, provides examples based on real cases, and offers guidance about incorporating responses to social needs into water management decisions.

Defining the Arena

The arena where sociopolitical needs are considered involves complex and interwoven issues that affect many aspects of society. To understand these, it is necessary to define several terms that seem related. This list explains them by synthesizing definitions from common sources, like dictionaries and government agencies.

- **Equity** is different than "equality" and means being fair and impartial to individuals within organizations or systems. It is sometimes expressed as "social equity" and used often in water resources management.
- **Equality** is the state of being equal, especially in status, rights, and opportunities. In public policy the term "equal opportunity" has been used. If groups received the same quantities of water, that would be an example of equality.
- **Ethics** are moral principles that govern behavior or conducting activities. This term occurs as a control on professional, business, and government practices.
- **Social welfare** means the well-being of people including physical, educational, emotional, mental, economic, and spiritual needs.
- **Social justice** is application of the concept of justice in the distribution of wealth, opportunities, and privileges.
- **Environmental justice** is an aspect of social justice that means fair treatment and meaningful involvement of all people regardless of status with respect to environmental conditions.

These concepts are interrelated and difficult to fit into fixed boxes. In the social issues of water resources management, all of them are in play. There may be tension between people's views of how they apply in water situations, but water managers should seek balanced solutions that are as neutral as possible within conflicting goals. This is challenging because the conflicting goals involve allocation of society's resources. Ultimately, some decisions are left to the political process and the judicial system.

Although water managers normally do not have authority for social welfare decisions, water programs can support positive approaches. This responsibility is recognized by the codes of ethics of professional groups. For example, civil and environmental engineers address social goals via the ASCE Code of Ethics. It requires engineers to protect the health, safety, and welfare of the public; enhance the quality of life for humanity; treat all persons with respect, dignity, and fairness, and reject all forms of discrimination and harassment; and sustainable development; balance societal, environmental, and economic impacts (ASCE).

Water Management and Social Welfare

The concept of social well-being can be used to focus the responsibility of water managers to respond to sociopolitical goals. Among the physical, educational, emotional, mental, economic, and spiritual needs of people, water management deals primarily with the physical and economic categories. More formally, when the role of water in meeting the hierarchy of human needs is considered, it addresses needs mostly at the basic levels (Figure 16.1). This hierarchy of human needs has become a popular psychological concept that expresses basic needs at the bottom and displays discretionary needs at higher levels. When applied to water management issues, the different uses of water are apparent.

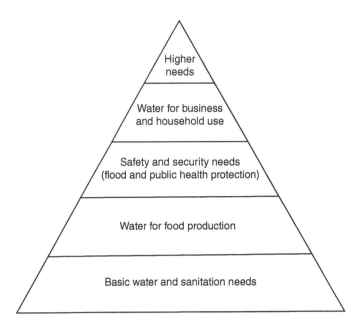

Figure 16.1 Hierarchy of needs for water.

Not surprisingly, water and sanitation needs are at the most basic level. The uses of water and water infrastructure transition to the top, which is labeled "higher needs" to express uses such as water for recreation and esthetics.

Social welfare is addressed in paradigms for water management. For example, the definition of IWRM includes to "maximise economic and social welfare in an equitable manner." The explanation is that water managers can promote social change and the changed behavior of water users toward sustainability and to foster water security. This changed behavior is about attitudes in people and institutions and considers that one person's benefits may be losses to others. Depriving social groups from participatory processes marginalizes poor or vulnerable people and favors more powerful groups (GWP). A companion paradigm, Total Water Management, includes in its definition to promote the greatest good of society by fostering public health, safety, and community good will (AWWA). The ASCE Code of Ethics mentioned earlier also requires consideration of social welfare.

The consensus of all the guidelines is that water managers should promote social welfare, but the challenge is to find specific and effective mechanisms to accomplish this. This challenge has been addressed continually, but has proved difficult to implement. The Water Resources Planning Act provided and the experiment, which will be described next.

Principles and Guidelines

The Water Resources Planning Act required development of a set of Principles and Standards to guide preparation of plans. These were prepared for economic, social, and environmental aspects, but the social aspects proved to be most controversial. As the programs under the Act were disassembled, the Principles and Standards were continued but modified in the form of a set of Principles and Guidelines. These included a discussion of "Other Social Effects," which recognized the need to consider them but stop short of definite guidelines. The discussion ranges across effects that included the quality of community life; life, health, and safety factor like flood risk; displacement of people, businesses, and farms; long-term productivity such as agricultural land; and energy requirements and energy conservation. Trying to consider these abstract notions leads to vague explanations, which are difficult to implement and require interpretation for each situation by water resources managers.

To identify social impacts of water management into guide responses, a framework model is needed to avoid the subject becoming too broad and vague. The model should ask the following questions: what do people need from water management and what can be supplied?

Meeting Social Needs Through Water Management

One way to approach the framework is to link it to the hierarchy of human needs that was discussed earlier. Flipping the hierarchy enables us to create a list of priorities with a view toward human needs. This list tends to track the purposes of water resources management and by paying careful attention to equity, justice, and social welfare, appropriate responses can be crafted.

Water issue	Human need
Drinking water	Safe drinking water for health and life
Sanitation	Access to safe sanitation and hygiene
Affordability	People can afford water services
Food production	Water for food production/irrigation
Security	Not being flooded, not displaced from homes due to projects
Water pollution	Not getting sick from environmental water
Resource access	Access to clean and abundant water in the environment

The categories shown are detailed enough to provide a framework for discussion, but they are not exhaustive because connections among social needs and water are so extensive. Responses to each issue will depend on context. Most of the categories are being addressed in the United States, although this can mask problems among sectors of the population and areas of the country. The situations in the low-income countries of the Global South include many situations with inequities and lack of attention to social justice. How the categories play out in different situations can be illustrated by discussions of the goals for each category and identification of gaps that should be addressed.

Access to safe drinking water is the signature global water-related social justice issue because without it, maintaining health and vitality is impossible. Such access is often asserted to be a human right. While the level of service that is a claimed human right can be debated, providing basic access to everyone seems fair. In the United States, access to safe drinking water is close to universal, but problems remain. Mainly, the issues revolve around affordability questions and whether discrimination is practiced in extending services. In low-income countries, many women must still carry water long distances daily, and in some countries people share water sources with animals. Whereas access to water provides abundant supplies in the United States, in many places globally quantities available are so low and unsafe that dehydration and poor hygiene are inevitable. Programs around the world to provide safe drinking water are those that address this social need. The evident response needed is to ensure safe drinking water to everyone with attention to the possibilities and limits imposed by the contexts, but ensuring at least access to meet basic human needs.

Access to toilet facilities and adequate water for personal and household hygiene are essential for society to function. In the United States, these are mostly available, but depend on income level. The problem of access for homeless people is acute everywhere. In low-income countries, access can be poor or nonexistent. In some schools in these countries, no water supply is provided for hygiene, and sanitation facilities are often missing, with especially severe impacts on females. An appropriate response to this human need is to provide adequate access to water and sanitation, especially to meet the needs of females.

While basic levels of water services are considered as a human right, how to pay for them is an enigma. It is undeniable that providing the infrastructure and operational capability to provide water services requires adequate financing. If all people cannot pay their share, the only way to provide the services is through subsidies of different kinds. In the United States, attention is being given to subsidies and affordability, but water rates are local matters and cannot be mandated by higher levels of government. In some low-income countries, inequitable situations are widespread. Evidence is that in some slum areas, people pay much more for water than higher income people in the same cities. Affordability is the difficult issue because it involves a transfer of income either through tax-supported subsidies or cross subsidies. A starting point to address the issue is to recognize its importance in search for the range of possibilities to ameliorate the problems.

The issue of equity in irrigation water applies mainly to smallholders who depend on it for income and food security. Problems occur in rural areas where water management is often missing. These can be caused by drought, lack of allocation of water, polluted water, and poor management of irrigation systems, especially the tailender problem where upstream diverters take all of the water. Mutual irrigation companies can provide for equity among water users, but addressing needs of disenfranchised people requires attention. The social issue of ensuring participation by all groups is especially evident when irrigation water is scarce and power is not distributed equitably. The water managers most involved here are local and pursuing equity in water access and distribution through mechanisms such as water user associations is a promising solution.

Flood risk control and stormwater management are needed to protect people from damage, injury, and death. In the United States, deliberate programs are underway at all three levels of government, but in low-income countries, people are often exposed to greater risks. Even in the United States, dam and levee safety are security threats where failure to maintain infrastructure can lead to disastrous consequences to vulnerable populations. For example, the failure of levees in New Orleans due to Hurricane Katrina in 2005 had massive impacts on many low-income people. Dam failure in low-income countries is serious as, for example, failure of a mine tailings dam in Brazil in 2015 which devastated communities and spread toxic wastes over hundreds of stream miles. Urban flooding is a systemic problem of cities in low-income countries, which face urbanization, social inequality, and environmental degradation. The threats stress the institutional capacity of cities, and they are unable to cope. Low-income people often live in flood-prone areas, do not understand the hazards, and lack institutional support. This problem is increasing in magnitude due to attractiveness of cities, migration, poverty, lack of community cohesion, and overwhelmed infrastructure and management systems. The social aspects of flood and stormwater problems are challenging because complete solutions are not forthcoming through direct government programs and collective action is required in local areas. This in turn requires effective social organization which can be challenging when conflicts are present among local groups.

Security is also threatened when people are displaced from their homes due to a water project such as in dam construction. If a community decides to locate a wastewater treatment plant in the low-income area, it can threaten people's security through potential exposure to pollution as well as through lower property values. This problem should be addressed through sensitive consideration of the impacts that water projects will have on different groups.

Environmental water quality is important for health and livelihood everywhere. Often low-income people who live in polluted rural or urban areas are exposed to health hazards. Even with modern public health systems, water-related threats come when people are exposed to contaminants in the environmental water. This can occur in irrigation, swimming, and during flood events. Contamination of water is a difficult problem to solve, especially in the case of nonpoint sources. Even point sources can create severe problems when wastewater treatment is not established or managed well. Equity and environmental justice are served when effective water quality management is achieved.

As a final category, access to water resources for fisheries, amenities, and potential spiritual values is important for all people. If this type of access is only provided for higher income people or people from particular groups, injustice is evident. Once again, this is a matter of power and privilege. Providing social benefits to low-income people is controversial because, once again, it involves a transfer of income. Fair and prosperous societies must develop collective ways to address this challenge.

How Can Water Managers Address Sociopolitical Goals?

Water managers can address sociopolitical goals through effective infrastructure and programs and by evaluating the equitable aspects of decisions. While they cannot decide issues of politics and equity, they can inform decision-makers about them and the trade-offs. To do this, they must be able to assess and explain the

Table 16.1 Human needs for water.

Human need	Water management responses
Safe drinking water for health and life	Ensure safe drinking water to everyone with consideration of the context, while ensuring at least access to meet basic human needs.
Access to safe sanitation and hygiene	Use available means to provide adequate access to water and sanitation, especially to meet the needs of females.
Affordable water services	Give priority to affordability and search for the range of possibilities to ameliorate the problems.
Water for food production	Pursuing equity in local water access and distribution through mechanisms such as water user associations.
Security from flooding and displacement	Promote effective social organization for mutual flood risk management. Ensure justice in all aspects of water projects, especially in impacts on the poor and vulnerable.
Water pollution	Ensure effective water quality management programs are achieved comprehensively.
Resource access	Develop collective ways to provide access to all benefits of water resources.

social aspects effectively. A starting point to make these assessments can be created from the hierarchy of needs and the examples of water needs that align with it.

Table 16.1 presents a generalized set of responses that can be used in specific situations to initiate social impact assessments. The contextual situations that occur across the many water resources management scenarios prevent us from making specific prescriptions. Water managers can derive these inappropriate ways by considering the suggested responses and applying them to specific situations.

In planning and problem solving situations, the water manager can identify the impacts of each water action or inaction. Then, the manager can assess if anticipated actions make each category better or worse, examine effects on cohorts of people by income, gender, age, ethnicity, or others, and create a matrix to indicate how any water action or lack of an action would impact the elements of the community.

The criticisms of water managers for not being sensitive enough to social concerns reflect divisions in society between privileged and less privileged groups. These divisions have many dimensions, such as human rights and safety nets in society. Closely associated is the question of responsibilities versus rights, which is reflected in the affordability issue. There are no easy answers to the questions that arise, but there are definite paths that we can take to promote social equity though water management actions.

Questions

1 Define the following terms and give an example of a water management situation involving each: equity, ethics, and environmental justice.

2 Explain what is meant by social welfare. How does it relate to environmental justice?

3 How were social goals addressed in the Principles and Guidelines of the Water Resources Planning Act?

4 Choose one of the examples of social goals listed here from the Principles and Guidelines and explain how it might be addressed through a water resources plan: employment distribution, especially the share to minorities; life, health, and safety; displacement of people, businesses, and farms; maintenance and enhancement of the productivity of resources, such as agricultural land.

5 Explain if and how the ASCE Code of Ethics addresses environmental justice.

6 Explain the criticism that experts in water resources management cannot be relied upon to address social issues adequately.

7 Give examples of how the following water issues relate to environmental justice: drinking water; sanitation; water for agriculture; security against flooding; clean streams; and drainage.

8 The Global Water Partnership stated that social change instruments are not neutral. Identify one social issue for water management and give an example where one group might gain while another lose through a water action.

9 How does public participation in water decisions relate to empowerment of marginalized populations? Give an example.

10 What is an example of a claimed human right to water and what might be a corresponding responsibility?

11 What are the main activities of the UN relating to human rights to water?

12 In what ways might the UN and the World Bank address environmental justice? Could financing institutions in the United States use the same methods?

13 What is the joint monitoring program of WHO and how might it address environmental justice?

14 In your opinion, what is the best way for people living in the United States to help with water issues in other countries?

15 Explain how IWRM can be a tool to address social issues.

16 What is meant by two world views colliding in IWRM?

17 What would be the social and economic theory behind the phrase in the IWRM definition "... to maximise economic and social welfare in an equitable manner ..."?

18 Formulate and explain a hierarchy of human needs for water and water services.

19 Give examples how social issues are involved with these nexuses: water–society; water–security; water–food; water–health; water–food–energy.

20 Give an example of a water amenity.

21 What is a social contract in the sense of water management?

22 Explain the evolution of the concept of water as a human right.

23 Give an example of a gender issue in water management.

24 What is social impact analysis? List categories of social impacts.

25 Give some examples of impacts on people of water infrastructure development.

26 How should equity and responsibility be balanced in water management?

27 Name several players in the international community with strong involvement in the social issues of water management.

28 Explain why social and environmental needs are not valued well by market choices.

29 How do social impact goals relate to sustainability, corporate social responsibility, and environmental, social, and governance (ESG) reporting.

30 Explain the social issues that may arise in planning for a contentious infrastructure project. Use the example of a wastewater treatment plant to illustrate.

17

Environmental Planning and Assessment

Introduction

Large-scale threats to environmental integrity are evident in metrics such as climate change, melting ice caps, declining biodiversity, and others. Reports show alarming drops in species populations during recent decades, with much of the loss caused by habitat destruction due to unsustainable practices in agriculture, logging, and other economic activity. These occur mostly at local levels, where water management is an integral part of most of the episodes. Planning for water management actions will increasingly include environmental assessments to identify and prevent further losses.

Halting environmental decline is a major challenge facing the world. Recent decades have seen increased requirements for environmental impact assessments to work on it, but full success has not been achieved. Environmental assessment requirements in the United States began with the 1969 National Environmental Policy Act (NEPA), which has driven many of the emerging methods. Environmental impact statements required by NEPA receive much attention due to their importance, costs, and perceived effect on delaying projects. They deal with diverse actions, including those involved with water. Implementing environmental programs at higher levels of government involves a process called Strategic Environmental Assessment (SEA). The concept is to embed environmental considerations in policies, plans, programs, and projects.

For water resources management, environmental impact assessment is an important process at all levels. A large project with federal requirements may have a EIS process that extends over years. At lower levels of government and small scales, environmental concerns about projects may fly under the radar screen but still be important to many citizens. The cumulative effect of the smaller actions is

Water Resources Management: Principles, Methods, and Tools, First Edition. Neil Grigg.
© 2023 John Wiley & Sons, Inc. Published 2023 by John Wiley & Sons, Inc.

likely more important than the large-scale projects because of the numbers. In many cases, local actions involve either federal or state controls, and environmental assessment processes may be like those for federal actions.

The purposes of this chapter are to review environmental impact assessment as a process and to illustrate how it can be used in water management to go along with economic and social assessment to form the third leg of the stool represented by the TBL concept. The chapter reviews the legislative and regulatory background, describes activities at the global level, outlines procedures under NEPA, and explains how project-level assessments should be done.

Environmental Indicators and Biological Diversity

Many factors are considered in environmental assessment, including biological diversity. The overall goal is to lay the foundation to sustain environmental integrity. Although environmental integrity has general goals about sustainability, it is multifaceted and there is no simple way to measure it.

Of the SDGs, the two goals that address environmental integrity most directly are Life Below Water and Life on Land. Both depend on water management to a great extent. Assessing progress on these is challenging because both are broad. Assessment tools have been made available, but they are mostly based on qualitative data.

One barrier to more quantitative environmental assessment is development of effective environmental indicators. USEPA's Report on the Environment is a leading publication about indicators (https://www.epa.gov/report-environment) and it provides more than 80 of them to answer 23 questions in five theme areas. The five theme areas are air, water, land, human exposure and health, and ecological condition. Each of these involves water issues to some extent. Ecological condition is closely related to water management, and questions address trends in ecological systems, diversity and biological balance, and their physical and chemical drivers.

In the case of the water theme area, the questions are:

- What are the trends in the extent and condition of fresh surface waters and their effects on human health and the environment?
- What are the trends in the extent and condition of ground water and their effects on human health and the environment?
- What are the trends in the extent and condition of wetlands and their effects on human health and the environment?
- What are the trends in the extent and condition of coastal waters and their effects on human health and the environment?
- What are the trends in the quality of drinking water and their effects on human health?

- What are the trends in the condition of recreational waters and their effects on human health and the environment?
- What are the trends in the condition of consumable fish and shellfish and their effects on human health?

The breadth of each of these questions indicates the complexity of the arenas to be assessed. For example, first question asks about trends and condition of waters and their effects on people and nature. Perhaps the question can be answered for a small-scale project, but for any situation involving higher scales, it seems certain that specific answers will be difficult to develop. That question has been assigned nine indicators that relate to fresh surface waters. One indicator presents information about stream flow patterns, which is an aspect of surface water extent. Another looks at the amount of water that the nation withdraws from surface water bodies every year. The other seven indicators characterize various aspects of condition, including the physical condition of sediments, the condition of biological communities, and the chemical condition of the water itself. One of the most significant indicators is about elevated nutrients in waters ranging from small streams to large rivers how they lead to problems offshore.

Global Level Focus on Biodiversity

As background it is helpful to know the evolution of the biodiversity problem and our understanding of it. During recent decades there have been many reports about biodiversity issues, and at the global level the focus has been on international conventions that address water issues, usually with other issues. An international convention is an agreement between countries about action in different areas, including environmental issues.

One of the earliest meetings led to the Ramsar Convention on Wetlands of International Importance in 1971, where the treaty was signed in Ramsar, Iran. Its implementation has led to an organization that sponsors periodic meetings and works with NGOs as international organization partners, including WWF and IWMI. The United Nations Environment Programme (UNEP) convened a working group in 1988 and it led to a Convention on Biological Diversity, which was signed in 1992 at the United Nations Conference on Environment and Development (Rio Earth Summit). At the global level, most activity is with the UN system and the Convention on Biological Diversity, which requests parties to the Convention to apply impact assessments to policies, plans, programs, and projects. The Secretariat issues periodic reports about progress toward a target set in 2010 in Aichi, Japan.

The Convention includes goals for maintaining freshwater ecosystems. For example, one goal set in 2010 was "By 2020, pollution, including from excess nutrients, has been brought to levels that are not detrimental to ecosystem function and biodiversity." This is an example of broad and ambitious goals, which must continually be sought so that progress can be made.

While US never signed the Convention due to political disagreements at home, it pursues similar environmental goals that seek to reduce biodiversity loss. These are implemented mainly through environmental laws and controls, such as the requirement for impact statements.

NEPA Background and Process

The United States started to take leadership in 1969 went past the National Environmental Policy Act. Some of the details of NEPA are explained in Chapter 20, which explains its basic provisions. NEPA establishes National Environmental Policy and administrative procedures, focuses on federal agency actions, creates a Council on Environmental Quality (CEQ) is a White House agency, and establishes environmental assessment requirements. With more than 50 years of history, the story of NEPA is documented thoroughly and details are available from USEPA websites and other sources.

NEPA's assessment requirements form a hierarchy that begins with categorical exclusions for actions without individual work cumulative impacts, then go the environmental assessment (EA) level which may find no significant impact (called a FONSI) or lead to the most stringent level, the full environmental impact statement (EIS). Although NEPA is aimed at federal actions, it affects private parties when there is some federal involvement in a private action such as permits or use of federal lands. According to USEPA, each year there are about 500–600 EISs and about 50 000 EIAs.

NEPA has extensive public involvement requirements, regardless of whether the funding is a FONSI or that an EIS is needed. After completion of an EIS and a corresponding decision, the lead agency prepares a Record of Decision (ROD). Since 1994 environmental justice requirements have been included and require federal agencies to assess factors that might have disproportionately high and adverse effects on human health or the environment especially considering minority and low income populations.

Guidelines for Environmental Assessment

Leadership in requiring environmental assessments for federal water projects stems back to the 1965 WRPA, which required Principles and Standards for planning that included assessment of environmental impacts. These have been replaced by the Principles, Requirements, and Guidelines, which require agencies to integrate their analysis into planning processes, in the same way the NEPA process and land management planning are integrated into larger planning processes.

Many state and local projects have federal links that require compliance with NEPA. The way the assessments are approached varies by state. A search of state environmental requirements for projects will lead to examples of processes and requirements in many states. For example, California has extensive requirements and has published a handbook about how project developers can integrate federal and state rules.

Tools for Environmental Assessment

In the early days of environmental assessment, the work required was laborious because project developers had to determine the structure of their reports, collect extensive data, and provide analyses without specific guidance. Currently, many tools have been developed to provide assistance in the planning process, although data requirements and analysis remain complex.

For example, USEPA has developed a tool called NEPAssist to facilitate environmental review processes by using data drawn from EPA GIS databases and the web to provide screening of indicators for an area of interest. It includes data related to impaired water points, streams, water bodies, and catchments, and on sole source aquifers as well as wild and scenic rivers. This is an example of using large databases to provide information for specific purposes.

For local projects, use of environmental and sustainability rating tools such as Envision has become common. Envision has 64 sustainability and resilience indicators called credits, which are organized in categories for quality of life, leadership, resource allocation, natural world, and climate and resilience. Each credit has levels of performance goals from low to high contributions to sustainability.

The Envision documentation claims that its criteria address economic, social, and environmental issues jointly, in categories such as human wellbeing, mobility, community development, economy, materials, energy, water, conservation, ecology, and emissions, among others.

Each of the indicators used in Envision is broad and spans multiple topics. The water parts of the tool address treatment, distribution, stormwater, and flood control and nutrient management. The credit list for water in the resources category shows the ratings for goals to preserve water resources, reduce operational water consumption, reduce construction water consumption, and monitor water systems. Not surprisingly, water criteria show up in other categories as well. For example, to manage stormwater is in the natural world category under conservation.

Points assigned for the various goals are broad and include achievement levels with ratings called improved, enhanced, superior, conserving, and restorative.

Questions

1   Outline a chronology to trace the development of environmental assessment requirements from the 1960s through today and include a discussion of US and international approaches.

2   Explain what NEPA is and how it works.

3   Give an example of a situation using environmental assessment.

4   Explain how the Envision rating tool can be used for TBL analysis of local water projects.

18

Economics of Water Resources Management

Introduction

This chapter and the next address how to use economic and financial concepts and tools to aid decision-making in water resources management. Economics and finance seem similar but have different applications in water decision-making. Economics explains criteria to aid decision-making about allocation of resources of different types, including water resources. Financial tools, which are discussed in the next chapter, are used to determine how to pay for the infrastructure and the programs that enable water organizations to deliver services. So, economics is more about concepts and analysis, and finance is more about how money is involved in water resources management. They share common calculation methods, such as interest rate computations.

The way that economic concepts and tools are used in water resources management creates a sub-discipline of water resources economics. Its tools support decisions that allocate society's water resources by maximizing benefits in the public interest. For the most part, market economics do not work when assessing economic choices related to water management, and public sector economics are required. For example, a common question about water allocation is how to price the resources to promote efficient use, even if the pricing scheme goes counter to consumer demand. As market methods do not work well, non-market methods are needed to incentivize efficient allocation of water and to assess benefits from water management.

Economics has a quantitative side, which in the case of water decisions is usually manifest as interest rate calculations to discount future values to the present and assure that decision support is provided on a level playing field by comparing apples to apples. It remains difficult, of course, to quantify many benefits and costs. Economic computations are explained to lay the foundation for a discussion of decision-making, especially benefit–cost analysis (BCA) and cost effectiveness analysis. This quantitative material draws on the discipline of financial analysis

Water Resources Management: Principles, Methods, and Tools, First Edition. Neil Grigg.

(sometimes called engineering economics) and on related tools to compare alternative choices.

In this chapter we focus on both the big picture of economic concepts and on the narrower topic of how to compute key economic quantities to aid in decision-making. The economic concepts show how water services are classified as public goods and how public and private management approaches are used, including government enterprises. Water services are usually provided by monopolies, so regulation is required. Investment in water infrastructure as an economic stimulus receives only brief mention because other infrastructures, especially transportation, have greater impacts on productivity. Social justice and water management are strongly linked as explained in Chapter 16. Sustainability in water resources management is also an important economic topic as explained in Chapter 17.

Water resources management is multi-objective, and the chapter also explains how to analyze trade-offs using the triple bottom line (TBL) concept and multi-criteria decision analysis (MCDA) tools. The TBL offers a conceptual framework to link economic, social, and environmental objectives and outcomes, and MCDA provides an organizational framework to consider the tradeoffs and uncertainties.

Economic Concepts Applied to Water Resources Management

Economic concepts explain the principles for allocating resources and solving related problems. To apply these requires understanding of the contexts, the possibilities, and the limitations of economic analysis. One definition of economics is that it is a social science that addresses the satisfaction of needs and wants by allocating scarce resources among alternative and often-competing uses. Water economics is therefore the study of scarcity and choice in the ways water can be used to satisfy demands when resources are limited and choices about allocations must be weighed. Allocation of water resources involves more than just water itself and extends to its many uses to satisfy water-related needs.

Global discussions about water, such as lack of water and sanitation and water scarcity, focus on economic topics such as the Sustainable Development Goals (SDGs), water justice, tenure, security, human rights, and issues in low-income countries. Discussion of these occurs regularly in global institutions with economic roles in water, such as the United Nations and World Bank. They also form important parts of national water plans such as those discussed in Chapter 13.

These broad topics create a framework for use of water resources economics to aid decision-making by maximizing benefits in the public interest. In addition to the analysis of monetary returns, the framework includes issues such as incentives to use water and how arrangements for sharing can be structured. As Figure 18.1 shows, water is a connector among other resources available to society, including

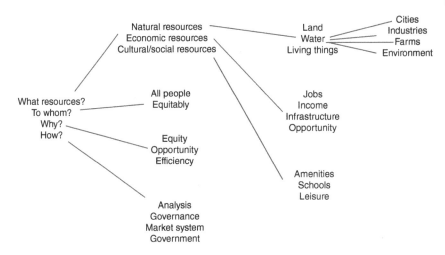

Figure 18.1 Scope of water resources economics.

natural and economic resources as well as resources provided by social and cultural systems. These resources should be allocated in the public interest on a sustainable basis to maximize social welfare. Water resources management is the mechanism to allocate water to cities, industries, farms, and the environment. Also, water resources management allocates resources like their use to convey wastewater and generate hydroelectricity. The allocation is to all people in the public interest and should be equitable, which is a broad concept generally addressed through law and politics. Wise allocation should promote economic advancement, opportunity, efficiency and use of resources, and equity.

Allocating water resources as wet water means to authorize use of the actual water itself. This requires measurement of satisfaction resulting from decisions about quantity, quality, and timing of water availability. Another aspect of water resources is the use of their management facilities, such as reservoirs and channels to address flood control or other related purposes. As mentioned, market economics are normally not effective in allocating either wet water or use of facilities, which creates market failure and requires public sector economic analysis. In particular, supply–demand theory does not provide the tools to manage water resources. While markets do not work in general, some of their tools, such as trades or auctions for the right to discharge pollutants to streams, can introduce a version of the market into water management. In planning, non-market values must normally be estimated, such as, for example, the value to society of the esthetics of a waterfall.

In most cases, allocation decisions focus on balancing among competing interests. For example, irrigators may want water for their crops, but others may want to leave it in the stream for fish and wildlife. Irrigation is mostly an economic goal, whereas water for fish and wildlife is mainly an environmental goal. As another example, you may want the water level in a reservoir to remain high to generate hydropower, but it should be low to capture flood waters. In the first case, you are allocating wet water, but in the second case you are allocating storage space in a reservoir.

As explained in Chapter 13, the main allocation problem in water resources management is how to foster successful voluntary collective action. As an example of the economic aspects of collective action, its failure can be seen in a community's unwillingness to voluntarily spend money to treat wastewater so a downstream community can have cleaner water supplies. At the same time, this upstream community may be willing to invest to provide good quality drinking water. Some authority must dictate to the upstream community the requirement to pay for and treat their wastewater.

Public and Private Economic Goods

Because there are different types of water uses and water services, the economic concept of public and private goods is helpful to create context for the different water uses and to suggest best ways to finance them. The concept is based on classifying economic goods by whether they are rivalrous (the uses compete with each other) and excludable (you can exclude someone from using them).

The four cases can be explained by a matrix of public and private goods (Figure 18.2). Water supply is an example of toll goods and can sold for a fee. Flood control is a public good. Water left in situ, whether for navigation, esthetics, recreation, or aquatic habitat, is a common pool resource. Water in a country club reservoir is a private good. The water in a reservoir may be a private or a toll good, but the storage capacity of the reservoir used for flood control is a public good.

If water resources are toll goods or private goods, they can be financed through rates and charges. If they are public goods, this is not feasible, and taxes are used to finance them. If they are common pool resources, measures must be taken to ration them as in the case of instream flows that are maintained, despite continuing heavy demands to divert the water.

	Excludable	Non-excludable
Rivalrous goods	Private goods (Private lake)	Common-pool goods (Instream flows)
Non-rivalrous	Toll or club goods (Water supply)	Public goods (Flood control)

Figure 18.2 Classification of public goods related to water.

The classification system of goods is not perfect, of course. Nobel Prize winner Elinor Ostrom proposed to modify the categories to consider incentives and the fact that both rivalry and exclusion can be variable and not absolute. For example, in the case of water quality we can say that stream assimilation capacity as a good is rivalrous because as it is used, its capacity is reduced. In the case of excludability, we might have to provide a water supply for customers who cannot pay in certain cases. Due to complexities like these, Ostrom found that water management was challenging to fit into her theories about management of common-pool resources.

The classification of public and private goods signals us as to which water services are more appropriate for public sector management and which might be offered by the private sector. Although water resources should be allocated in the public interest, some water services can be provided by private companies rather than through government enterprises like city governments or special authorities. When government-owned facilities are sold to the private sector it is called privatization. Contracting out services like operation of a treatment plant is sometimes also called privatization. A more recent popular term for this is public–private partnerships (PPP).

Providing water services inherently introduces monopolies which require regulation. It is not in the public interest to allow competing organizations to raise the cost of water services to the public based on private interests. Private water companies are subject to rate regulation by state commissions, although government-owned water utilities are only subject to political regulation by their governing boards. Most activities by federal and state agencies are normally not subjected to regulation but they are overseen by the political process and by the courts.

Economic Assessment

A common economics problem in water resources management is assessment of projects or other actions, considering both efficiency and equity, when mean contributions to wealth and distributions among stakeholder groups and regions. Contributions are measured in terms of benefits from the projects and actions.

Benefits include economic metrics like any gain in income or wealth. They can be tangible or intangible and direct or indirect. Tangible benefits are those you can measure, like an increase in profit from an investment. An intangible benefit cannot be measured directly, like an increased sense of security because a dam was strengthened and is less likely to fail. A direct benefit is one that stems directly from the purpose of a project, such as to increase regional income by using water to attract new jobs. To compare benefits from proposed actions, you must value them with some measure of relative merit like dollars. Benefits are measured by dollars produced in categories of benefits. Once they are adjusted to the same time basis and made commensurate, they are usually compared per the benefit–cost ratio or by net benefits. Categories of benefits in a water project might come, for example, from municipal and industrial water supply, agriculture, urban flood damage, hydropower, navigation, recreation, and commercial fishing.

Economic development effects of water infrastructure investments are often under scrutiny. This generally refers to the creation of wealth and income that grow the economy and enhance prosperity and quality of life. Water investments bring social, public health, and environmental benefits, but they may not stimulate much economic activity unless water was not available and by providing it the economic activity is enabled.

One example of economic development from water projects is creation of the Tennessee Valley Authority and the water development for power and flood control that it created to lead to jobs and stimulate the regional economy. Another example is the provision of water supplies to Southern California, which could not have grown to the extent that it has without the water. Conversely, consider a community that lacks safe, reliable, and affordable water and by providing water, the quality of life can be increased. This may lead indirectly to economic activity, but the water investment would not be the primary driver.

Social and environmental goals are inherent in water resources economics, which addresses environmental issues and concepts of equity, serving the public interest, justice, and links of water to other societal goals. Criteria such as affordability, need for subsidies, and rights to safe and reliable water supplies are basic considerations in water resources management decisions. Examples of water-related social issues are found frequently in allocation of irrigation water, pollution control, and flood and stormwater services, among others. Whereas economic analysis is performed quantitatively, social and environmental assessments rely more on qualitative information. Social impacts are especially hard to measure because different value systems are involved and constituents have competing interests. Environmental assessment is explained in Chapter 20 and social impact assessment is discussed in Chapter 17.

Multiple Objectives for Water Infrastructure Investments

Economic assessments in water infrastructure planning are inherently multi-objective in nature. As discussed in Chapter 14, economic analysis is the core of assessing project feasibility as part of decision analysis. The overall goal is optimum allocation of water resources among competing objectives. The analysis space is based on welfare economics with sustainability being a surrogate measure for optimum approaches. Economic tools and methods enable us to quantify and display parameters such as costs, benefits, and other metrics to support decision-making. Figure 18.3 outlines the methods and tools.

The TBL concept provides a framework to consider economic, social, and environmental goals simultaneously by providing a display of achievements and setbacks in the three categories. While it can include financial data, the economic

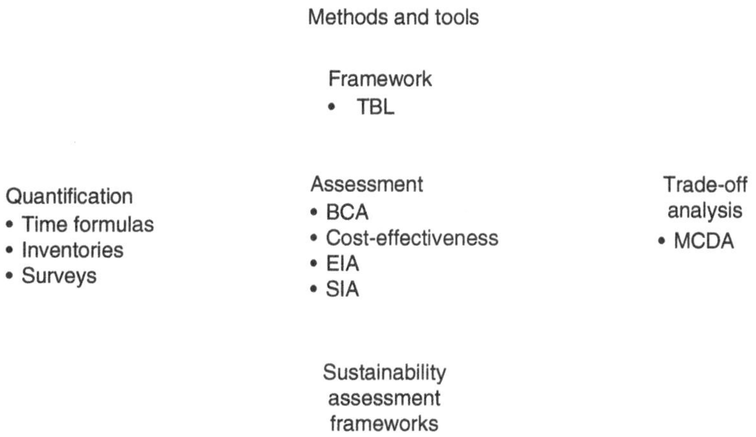

Methods and tools

Framework
- TBL

Quantification	Assessment	Trade-off
• Time formulas	• BCA	analysis
• Inventories	• Cost-effectiveness	• MCDA
• Surveys	• EIA	
	• SIA	

Sustainability
assessment
frameworks

Figure 18.3 Methods and tools of water resources economics.

category would address broad issues, such as economic development. Environmental and social accounts address positives and negatives for habitat, society, and related issues. A TBL analysis can be used in planning and as an addition to common financial reports.

To support the TBL, assessment methodologies are needed for economic, environmental, and social accounts. Quantification tools such as time value of money, environmental inventories, and social surveys can be used to provide the necessary information. Computer-based sustainability assessment frameworks are available to manage the information, see Chapter 17.

The basic TBL evaluation framework was illustrated in Chapter 13 as Table 13.1. It is basically a matrix where you can add positive and negative effects for economic, social, and environmental impacts of projects are they affect different

Table 18.1 Example of MCDA with weighting factors.

	Economic		Environmental		Social		Total weighted score
	Weight = 5		Weight = 3		Weight = 2		
	Score	Weighted score	Score	Weighted score	Score	Weighted score	
Project *A*	30	150	15	45	70	140	335
Project *B*	60	300	20	60	10	20	380
Project *C*	10	50	65	195	20	40	285
Totals	100	500	100	300	100	200	1000

groups of stakeholders. The matrix is amenable to MCDA analysis. MCDA is like a tool that can be used to consider what makes the best decision considering economic, social, and environmental goals through the TBL or other objectives established by the decision makers.

MCDA evolved from multiple strands of decision methods and is well-established as a common tool. In its use for economic, environmental, and social impact analysis, the concept is based on maximization of public welfare by seeking the best value of a "social welfare function," which includes categories of public goods such as economic development, environmental quality, or improved quality of life. MCDA measures how different strategies or projects lead to achievement in different categories of goals. While this seems straightforward and involves numerical data to measure outcomes and preferences, people have different opinions about what the numbers should be. For this reason, the outcome of an MCDA analysis is usually considered to be only one solution of many possible ones, and decision makers want to see how outcomes change as a function of sensitivities in the assumptions.

In a simple form, an MCDA display shows how strategies or projects score in the goal categories, as shown on Table 18.1, which is a modification of Table 13.1 Weights are assigned to the economic, environmental, and social goals. These goals could be stated more explicitly in ways that are appropriate for different situations. The project scores would be assigned after appropriate assessments, and by multiplication the weighted scores are determined. This yields a total weighted score for each project. In this simple hypothetical example, Project B has the highest weighted score. If the weights of the goals were changed, results would be different. If social goals were deemed to have higher priority, then Project A would score much better, for example. Economic scores can often be computed, but environmental and social outcomes are harder to quantify and often rely on verbiage rather than numerical scores to describe positive and negative impacts.

MCDA can be a useful tool for trade-off analysis. As is evident from the table, the projects could be modified to change the scores and negotiations could be conducted to change the weighting factors so that stakeholders could compare the choices from different perspectives and arrive at a decision that might be more satisfactory to a broad set of them.

Economic Assessment Considering the Time Value of Money

Performing quantitative economic assessments requires that we compare alternative choices, usually through some version of benefit–cost analysis. This requires computation of the time value of money to assess benefits and costs on commensurate timescales by using appropriate interest rates. Most computations involve applying interest rate formulas with different payment schedules. The methods can be complex, but most situations require only relatively simple computations involving cash flow diagrams and application of basic formulas.

The time value of money shows changes in value over time when money is either accruing interest or changing value due to inflation. You can adjust for the effects of inflation by changing all values to current dollars. The basis for the payment of interest is that there should be a return on capital, like interest paid on deposits in a savings account. For example, if you deposit $100 in a savings account at 2% interest, in a year you earn $2 and your savings are now worth $102. So, $102 in a year is equivalent to $100 at the present, given a 2% interest rate. Most computations of the time value of money involve applying interest rate formulas with different payment schedules.

Cash flow diagrams are useful to illustrate benefits and costs that occur at different times. Three variables are normally displayed: present value, future value, and annual value. For example, Figure 18.4 illustrates an original amount and the annual payments for a loan to be repaid with equal payments over time. How to use these will be explained by numerical examples using interest rate formulas.

Interest rate formulas have been used in engineering economics and financial analysis for a long time to compute payments that are equivalent to each other. In the past, the analyst would compute these or look up values in tables, but now use of business calculators, spreadsheet formulas, or other software makes computation more rapid.

Usually, six formulas are presented to show all situations you might want to calculate. However, you only need two main formulas, and the other four are readily derived from them. The first of the main formulas is the single payment compound amount factor, which shows a future value (F) for a present sum (P) compounded over n years at interest rate i. Shorthand notation is ($F/P, i, n$):

$$\frac{F}{p} = (1+i)^n$$

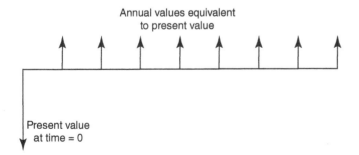

Figure 18.4 Cash flow diagram showing annual payments equivalent to present value.

where F is the future amount, P is the present amount, i is the interest rate, and n is the number of periods. For example, if a utility deposits $5.0 million into a reserve fund and earns 4% per year, the deposit will be worth $7.40 million in 10 years.

$$F = \$5.0^* (1.04)^{10} = \$7.40\,\text{million}$$

By using this formula for successive years, you can create a compound interest table or graph.

The second main formula is the capital recovery factor (A/P, $i\%$, N), which is used to compute a series of equal annual payments (A):

$$A = P \frac{i(1+i)^n}{(1+i)^n - 1}$$

For example, if a small town borrows $10.0 million from a fund to build a treatment plant and repays it in 15 years at 3% interest, the amount due at the end of each year is $837 665.80.

$$A = \$10.0^* [0.03^* 1.03^{15} / (1.03^{15} - 1] = \$837\,665.80$$

Most situations can be analyzed by combining these two formulas. For example, if the cash flow diagram showed a future payment of $10 million after five years and annual payments of $1.0 million, both at 5% interest, the equivalent total payment on the basis of present value will be:

$$\text{Total payment at the present time} = \$10\,\text{million} / (1.05^5)$$
$$+ \$1\,\text{million}([1.05^5 - 1] / [0.05^* 1.05^5])$$
$$= \$12.16\,\text{million}$$

These formulas are relatively simple to apply. Here are a few examples.
Future or present value of a single payment:
On day 1, you put $500 into an investment account that yields a nominal return of 7% per year. What is it worth in 10 years with monthly compounding?

$$F = 500^* (1 + 0.07 / 12)^{120} = \$1004.83 \text{ (note that 10 years is near the doubling}$$
$$\text{time for a growth rate of 7\%)}$$

You compute the present value of a single future value payment by inverting the same formula.
Future or present value of a uniform series of payments:
Now you put $500 per month (at the end of each month) into the investment account that yields a nominal return of 7% per year. What is it worth in 10 years?

$$F = 500^* (F / A, i\%, n) = \$86\,542.40$$

If the problem was to find what this future sum required in monthly payments, you simply invert the F/A formula to find A/F.

If you choose to compute the present value of this investment, the computation is:

$$P = 500^* (P/A, i, n) = \$43063.18$$

This sum, converted to a future value, is:

$$F = P^* (F/P, i, n) = 43063.18^* (1+i)^n = 43063.18^* (1+0.07/12)^{120} = \$86542.40$$

Note that the sinking fund, or the amount you need to "sink" into an account to pay for something in the future, is simply $(A/F, i, n)$.

Benefit–Cost Analysis

While economic assessment can be complex, relatively simple methods are often useful for assessing projects. Two approaches are frequently used: benefit–cost analysis (BCA) to compare the benefits (gains) of projects to their costs (losses) and cost effectiveness analysis to find the lowest cost solution to achieve a given goal.

Benefit–cost analysis is used to compare projects on the basis of their economic merits. As a method for water projects, the popularity of BCA emerged with the 1936 Flood Control Act, which required that projects could be authorized only if benefits exceeded costs, regardless of who they accrued to. BCA's concepts had evolved from earlier economic thinking and stem back a least to the previous century. This New Deal concept sought to provide a mechanism of public investment to benefit the nation at a time of economic hardship. BCA applies to problems of the public sector because it is a flexible procedure, and you can consider different categories of benefits and costs. It goes well with multi-objective analysis when you are considering environmental and social costs.

It was up to the Corps of Engineers to figure out how to implement BCA after the 1936 act, and detailed procedures have been developed for federal water projects. BCA has been used less for local water projects which are usually planned on the basis of financial feasibility rather than economic theory.

BCA requires attributes that include clear alternatives, merits of all appropriate alternatives, expected future consequences, viewpoint for weighing merits, differences among alternatives, non-monetary consequences (intangibles), and side effects. These apply across the board for decision-making, whether in private or public sector decisions.

As discussed earlier, benefits and costs can be tangible or intangible and direct or indirect. Tangible benefits can be measured but intangible benefits cannot be measured directly. A direct benefit stems directly from the purpose of a project, and an indirect benefit may occur to systems not related to the project, such as if a navigation project intended to reduce travel time also raised the value of adjacent properties. Given the approximations that arise in using BCA, most situations only involve tangible benefits and costs.

A water project will normally involve benefits that vary by purpose:

- Damages avoided through flood control
- Production added through irrigation
- Power generated
- Water supply sold
- Wastewater treated
- Water recreation created
- Navigation savings

To compare benefits and costs due to proposed actions, you value them in dollars. Once the valuations are made commensurate on the same time basis, the proposed actions are compared on the basis of the benefit–cost ratio or by net benefits.

To formulate alternatives, you would package mutually exclusive projects (Plan A, Plan B, etc.) or compile combinations of alternatives. The "Do Nothing" alternative should be included to create a basis for comparison.

You must specify the interest rate as a minimum attractive rate of return. This is sometimes difficult, but you may use the cost of borrowing money or an approved rate, such as one published by the federal government. BCA results are sensitive to selecting the proper interest or discount rate. Federal projects require use of a social discount rate, and guidelines are set by the Office of Management and Budget and explained in the Principles and Guidelines (see Chapter 13).

In cost-effectiveness analysis, a fixed requirement must be met at minimum cost. As an example of cost-effectiveness analysis, consider a wastewater treatment plant that must provide a certain level of pollutant removal. The required treatment level will guide planning and the solution will seek the minimum total cost. This is a version of comparing benefits to cost, but the benefits are fixed by the requirement.

Flood control is amenable to a BCA approach because a project can have variable benefits in terms of damages avoided. Recreation and navigation projects are also amenable to use of BCA, but in recreation projects, non-market estimates of the value of visitor-days are used. Navigation can be assessed based on savings in transportation costs. Irrigation projects are complex because the economic benefits may be low when food prices are considered, and social cohesion and food security must also be considered. A purely quantitative analysis might not be adequate. Water supply and power projects involve sale of commodities, and BCA analyses are hard to understand and not likely to be used for local projects. Wastewater projects are more amenable to cost-effectiveness analysis than to BCA because regulatory requirements must be met.

An example of comparing alternative public sector investments would be one water resources project versus another. Say you have a project that will return benefits of $1 million/year over its lifetime of 20 years, and you can achieve these benefits with a present investment of $8 million and annual operating costs of $200 000. In this case, the benefits are the annual $1 million returns. The costs are the initial capital cost of $8 million and the annual operating costs of $200 000. If the interest rate is 7%, then the computations are (using the formulas or a spreadsheet):

	Data	PV	AV
Investment	8 000 000	8 000 000	755 143
Operating costs	200 000	2 118 803	200 000
Returns	1 000 000	10 594 014	1 000 000
Interest rate	0.07		
Years	20		
Net benefits		475 211	44 857
BCR		1.05	1.05

These results were quickly generated using an Excel spreadsheet.

This analysis also shows how you can compute the rate of return (ROR). You would insert different interest rates until the BCR = 0. Using a trial and error method, this turns out to be 7.75%. Using the Excel IRR function, the answer comes back as 7.7547%.

Economic studies require consideration of inflation, which can be very significant. While inflation has impacts on the economy, its main impact is loss of purchasing power of money as measured by price indices. Say you build a project that costs $1 000 000 today. Inflation is at 4% per year. If you delay your project, inflation erodes the purchasing power of your money as shown in this table:

Year	Value ($)
0	1 000 000
1	961 538
2	924 556
3	888 996
4	854 804
5	821 927

So, if you delay until after year 1 you will not have enough money to do the project unless, of course, you are able to invest your $1 000 000 in an account that compounds interest so you can keep up with the inflation loss. If you can earn more money than inflation, you can even build up your funds and build a bigger project. That is why the advice you get is to borrow and build your project bigger now if you expect inflation to be high.

Questions

1 What is the attribute of water management that makes collective action imperative in most situations?

2 Explain the story of the "Tragedy of the Commons" to illustrate a failure in collective action.

3 How does "water resources planning" promote collective action?

4 The core question of economics is how a society should allocate its limited resources. In the case of water management, what are the resources to allocate? Try to include all categories you can think of.

5 Why are development banks, sometimes called infrastructure banks, needed to promote water resources management? Explain how they function.

6 Explain the concept of market failure as it occurs in water resources management.

7 Identify and explain one example of an "economy of scale" in the provision of water services.

8 What is meant by a water service organization being managed on an "enterprise basis?" Is this an example of free enterprise or is it a government mechanism?

9 Is provision of water supply to homes a public or private good?

10 Can water services be provided by for-profit organizations? If so, under what conditions can this arrangement work successfully?

11 What is meant by cost of service in rate-setting for water supply?

12 Should the price of water service be set at the average price for a user, or should it be set at the marginal cost of providing the next unit? Explain your answer.

13 What is the economic basis for providing a tax exemption for a water infrastructure bond?

14 What is a subsidy? Give one example where you think one might be appropriate in water management.

15 What is meant by equity in the allocation of water and water services? Give an example.

16 Explain what a monopoly is in the provision of water services.

17 How are water rates regulated in the cases of government and for-profit utilities?

18 In the case of wastewater services, how can downstream water users be protected from pollution by upstream dischargers?

19 What does a "human right" to water mean? Identify any constraints on this concept that must be faced.

20 Give an example of how decisions about water for irrigation might involve social cohesion and food security in a community, as well as quantifiable dollar benefits.

21 Give examples of social and environmental justice in situations involving irrigation water allocation to smallholders and in preventing water pollution.

22 If an irrigation farmer owns shares in a mutual ditch company and a growing city wants to buy the shares from him, what personal, financial, and social factors might the farmer consider in making a decision?

23 Formulate an example to explain a situation where MCDA can be used for trade-off analysis.

24 How does the TBL concept relate to the overall goal of improving social welfare?

25 How did benefit–cost analysis come to be used in the planning of federal water projects?

26 Is BCA normally used to plan local water projects? Why or why not?

27 Explain the application of cost-effectiveness analysis to a common water resources decision problem.

28 To prepare a BCA for a flood control project, what is the quantifiable benefit to be sought?

29 Why is it difficult to define benefits of stormwater projects in cities?

30 What is meant by "commensurate values" in application of BCA? Why is using them important?

31 In market economics, supplies of commodities are provided at attractive prices to respond to demands. Does this work for drinking water supply? Explain.

Problems

A Well to Augment Water Rights

You will develop a well to pump water to a river and satisfy surface water rights. The pumping enables you to deliver water during low flows and the depletion occurs during high flow so you can achieve a net gain. Your project would add 50 acre-feet/year with a favorable priority of water rights. The system would cost $100 000 for construction and equipment. The infrastructure and equipment would last 30 years and then require renewal. Consider reconstruction at 30- and 60-years from now and assume that it would be retired at year 90. Your interest rate is 0.04. O&M costs are $10 per AF per year. The water rights are currently selling for $5000 per acre-foot on the water market. *Calculate the present value of your profit from* this investment.

Reservoir

You will build a water project to store 50 000 acre-feet of water and yield an average of 10 000 acre-feet/year. If you finance the project by selling water rights as shares and assume responsibility for O&M, how much would you charge per acre-foot? The project cost is $20 million, and O&M will cost $1 million/year. The project will require an investment of $10 million every 20 years for renewal and have an infinite life. Ignore reinvestment costs after 60 years. Interest rate is 5%.

Irrigation

You are considering purchase of 500 acre-feet in perpetual water rights to irrigate 160 acres of corn. The land yields 150 bushels/acre if it is irrigated, but you can raise 75 bushels if it is rain-fed. The sale price of corn is $5 per bushel. Water rights sell for $3000 per acre-foot. What is your rate of return on using your water this way? Assume all other costs are the same, whether irrigated or rain-fed.

Hydroelectricity

You consider building a small hydroelectric plant that is rated at 1000 KW and can operate for 50% of the time. The construction cost is $5 million and the annual O&M cost is $250 000. The project must be rebuilt every 30 years. The interest rate you will have to pay to borrow money is 6%. What must you price each KWH at to break even? Solve it for the time horizon of 30 years without considering any rebuilding cost and then solve it for a 90-year total time horizon, with rebuilding at 30 and 60 years and decommissioning at 90 years.

Benefit–Cost Analysis

A flood control reservoir will cost $50 million to construct. Annual O&M cost will be $5 million. It will reduce annual flood damages by $8 million. The planning period is 50-years and the interest rate is 5%. What is the benefit–cost ratio of the project?

Plant Investment Fee

You are designing a new city for a population of 100 000 people. The water law is based on water rights, and you decide to buy 25 000 acre-feet to supply the city. This is enough water to be sure that in a dry year you will have enough to meet average demands. The cost to buy a water right is $10 000 per acre-foot. You must develop infrastructure for the water supply system, which will cost $5000 per capita to build. The annual cost to maintain the water rights and the infrastructure is $100 per capita. If the interest rate is 5.0% and the lifetime of the infrastructure is 40 years, what will be the total annual cost per capita for the utility to recover its cost? The lifetime of the water rights is infinite.

This is more water than will be needed in an average year when the use is 120 gpcd. The remaining water is not needed for the city in the average year. Assume this extra water can be leased out to farmers at $100 per acre-foot. Compute the volume of water in acre-feet not needed in an average year, the water rate per 1000 gal needed to break even if the extra water is not leased out, and the water rate per 1000 gal if the extra water is leased out.

Multi-Criteria Decision Analysis

Assume you are comparing three projects as shown. Notice that Project B scores highest with the existing weightings. Change the weightings so the social goal has a weight of five and the economics goal a weight of two. Recompute a new total weighted score.

	Economic		Environmental		Social		Total weighted score
	Weight = 5		Weight = 3		Weight = 2		
	Score	Weighted score	Score	Weighted score	Score	Weighted score	
Project A	30	150	15	45	70	140	335
Project B	60	300	20	60	10	20	380
Project C	10	50	65	195	20	40	285
Totals	100	500	100	300	100	200	1000

Energy Generation

A reservoir has a constant water surface elevation of 100-feet above a turbine, which operates at 80% efficiency. The average flow is 50 cfs during 12 hours of each day. Compute the KW/h of energy generated during one year of operation and the revenue earned if 1 KWH sells at wholesale for $0.03.

Loan Amortization

Prepare a table showing 10 years of payments on a 10-year loan of $1 000 000 at annual interest = 5%. Include all information shown on the following sample table.

Year	Loan payment	Loan interest	Principal payment	Balance remaining
0				1 000 000
1				
2				

Revise the table with a new calculation that results from an extra payment of $200000 at the end of year 3.

Tax Exempt Infrastructure Bond

Compute the net annual return for an investor who will buy either a 10-year tax-exempt infrastructure bond at 4% or a corporate bond subject to federal income tax but paying 6%. The investment will be $100000 and the investor is in the 35% tax bracket.

Loan Versus Bond Comparison

Prepare a table to explain the differences in cash flows and explain the difference in payback of the loan compared to a $1 million revenue bond with a coupon rate of 5%. Both have terms of 10 years.

Year	Loan payment	Loan balance remaining	Bond payment	Bond balance remaining
0		1 000 000		1 000 000
1				
2				

Infrastructure Bank

An infrastructure bank will loan $5 million to a water utility at a subsidized rate of 3%. The infrastructure bank must raise its funds by selling tax exempt bonds on the open market at 4%. The term of the loan to the utility is five years. How much money must the bank lay out over the five years to cover this subsidy?

Capitalize Maintenance

A roads department wants to capitalize annual maintenance fees of $5 million. It applies to the central government for a grant to fund this capitalization, at 5%. What must be the size of the grant to do this?

Depreciation and Deferred Maintenance

An infrastructure system that cost $100 million to build is estimated to depreciate 100% on a straight line basis over 40 years. What will be the book value at the end of year 30?

19

Financing Water Systems and Programs

Introduction

Water managers understand the importance of finance in their operations, and this chapter explains principles and ways to finance and provide funding for water infrastructure, services, and organizations. Financing refers to mechanisms like budget appropriations and loans, while funding means providing money for operations and capital expenditures. To illustrate, the utility might decide to *finance* a new treatment plant with *funding* from a revolving loan program. The term financing will be used in this chapter to mean both concepts.

A water manager working for a public agency or utility should know how to apply the principles of public finance to budget and account for money, raise funds for capital expenses, and set rates and charges. The manager should know the framework of financial management, but is not expected to be an expert in the details of accounting. The main goal of a water manager is to see that systems are funded well enough to meet their purposes and that management of funds is effective. Consultants need similar knowledge so they can advise their clients about financial issues, especially small governments and water utilities.

Financial issues are especially acute in many small water systems, which results in low organizational and management capacity and difficulty in financing infrastructure. This occurs because the financial base of small communities and rural areas is often low and the lack of management capacity may lead to inadequate rate structures and financial management problems.

To address these issues, the water manager and consultant should understand how finance and funding work at the three levels of government. This is addressed by the field of public finance, which is the study of the income and expenses of government entities with different purposes and scopes. Water finance involves incomes and expenses for operations at the federal, state, and local government levels. Public finance is different from private finance in that it focuses on public

Water Resources Management: Principles, Methods, and Tools, First Edition. Neil Grigg.
© 2023 John Wiley & Sons, Inc. Published 2023 by John Wiley & Sons, Inc.

sector issues, while private finance for individuals and businesses focuses on personal expenditures and the success of businesses.

At the federal level, operations of the USACE, the USBR, and other agencies such as USGS must be financed and funded. Each of these agencies has developed its own mixture of revenues, as, for example, the cost sharing of USGS through its cooperative stream gauging program. In the cases of the USACE and USBR, budgets must cover expenditures for workforces and to operate and maintain water infrastructures. Most state governments operate regulatory agencies with smaller budgets and do not own and operate large infrastructures. Depending on the function of the agency, funding may be primarily through the state budget or may include federal grants. At the local level, most water activities and delivery of services are conducted by water supply, wastewater, and stormwater utilities. Their financing and funding arrangements will be explained later as examples.

Financial Framework for Water Management

The general arrangements for water financing stem from the public nature of water as a resource, a required service, and an environmental element requiring management. Ultimately, water systems are financed through tariffs, taxes, or transfers (TTT). Tariffs involve charges for water use, fees for growth, and other charges. Taxes are sometimes used, but when water can be sold like a commodity, taxes are often not used. Transfers from higher government to lower government levels involve grants and special situations. Debt instruments like bonds are used to finance capital needs.

While some private water companies are in operation, they are regulated monopolies and perform the same functions as public water utilities. Differences in the ways that public and private water organizations manage funding will be discussed in a subsequent section about private water companies. All water organizations follow standard financial reporting principles outlined by the Financial Accounting Standards Board (FASB), and those in the public sector also follow public finance rules of the Government Accounting Standards Board (GASB).

Federal financial controls involve the Department of the Treasury, the Internal Revenue Service, and the Office of Management and Budget (OMB), among other agencies. State and local governments have smaller units that perform the same financial control functions. Federal water agencies are mostly supported by tax revenues, mainly from income taxes. State government agencies obtain revenues from taxes, fees, and transfers from the federal government. Sometimes, legislatures will shift the cost of government operation from taxes and transfers to fees.

Most funding goes into utilities at the local level, which operate as enterprise funds where revenue should meet expenses without subsidies.

A government enterprise that provides services can be a department within a general-purpose government or a stand-alone organization, like a public authority. At the US national level, the Tennessee Valley Authority is a large-scale example of a government enterprise. A state level enterprise might be a state toll road authority. At the local level, a city water department is like an enterprise that operates like a separate authority except for its governance structure within the city. In developing countries, government enterprises are sometimes called state companies or enterprises. For example, in Brazil, the water company for the State of Ceará (COGERH) is state owned.

Whether the enterprise is free-standing or part of a government unit, it normally follows the "enterprise principle" of management, and it does not depend on subsidies from general purpose taxes. This principle operates widely in the United States, but it might not be followed so closely by state-owned companies in developing countries. The enterprise concept is that services should be self-supporting and charged according to the benefits users receive from the services. Pricing through user charges is the basis for the control of the allocation of the services and for raising revenue. If a service is self-supporting, revenue generation and financial control are under the control of the manager rather than the political process. However, policies of governance authorities and desires of customers must be factored into decision-making.

Equity is also important in the enterprise philosophy and charging schemes should be fair. The case of providing vitally needed services when customers cannot pay the full cost of service is common. It can take a balancing act to offer low rates to such customers needing help while charging more to others, which is a cross-subsidy. Generally, subsidies are to be avoided, but equity may require them. The use of public subsidies for transit is common, for example, since the farebox does not pay the full bill, but this approach is less common for water services. Other examples of subsidies are in the now-defunct construction grants program for wastewater and the construction and operation of local public housing. The use of subsidies in developing countries is widespread, often providing the difference between life and death. In the case of irrigation systems in developing countries, for example, even though operation is not directly subsidized, it is indirectly subsidized through the allowance of deferred maintenance, with grants and soft loans for rehabilitation.

Water management involves both enterprises and tax-supported general operations, such as the case of government-owned dams. Enterprises should be funded by fees, while other operations may involve taxes and transfers. Water organizations can also incur debt to finance capital investments, but it must be repaid through TTT sources. Ultimately, all of the funds are from individuals, either through taxes or tariffs. Transfers from government, like federal grants, show how government does not create wealth but is a transfer mechanism that collects taxes from economic activity and fees for services and sends the funds back to individuals and public organizations.

Water utilities are capital-intensive and fixed costs like debt repayments or those that remain constant regardless of service level are normally high relative to variable costs like chemicals, maintenance, and training.

In most cases, tariffs and charges are monthly water bills and charges such as connections for new homes. Property taxes and sales taxes are used in some cases. Transfers are mainly grants from the federal or state governments or, in some situations, from donors.

Benefits provided by water systems to justify the funds they use are private economic goods like residential water supply and public economic goods like flood control. These goods provide essential support for health and welfare and they support the economy, but water services normally do not have a multiplier effect on economic activity. An exception is when no water services were available and providing them makes economic activity possible.

Plans and Budgets

The general approach to plan and administer funding for water systems is through some variation of a Planning-Programming-Budgeting System (PPBS). Estimates of funding needed for operations and capital are made through joint activities among water managers and finance officers. A Capital Improvement Plan will be used as a budgeting and financial tool to establish asset rehabilitation and maintenance priorities and funding.

Most water organizations have one budget for capital and another for operations. In the operating budget the details of ongoing expenses and revenues are projected, approved, and reported. Examples would be personnel costs, fuel, rent, and other recurring expenses. Capital budgeting considers longer time spans for items such as facilities and equipment. Construction projects, equipment, and acquisition of real property are financed through capital budgets.

Financing mechanisms differ for operating expenses (OPEX) and capital expenses (CAPEX). Customer charges through rates and fees are generally used for operations, while debt, savings, transfers through grants, and use of reserve funds are commonly used for capital investments.

Both operating and the capital budgets are planned on multiyear cycles. The capital budget is linked to comprehensive planning and needs assessment processes. The operating budget is linked to plans for services, organizational development and programs. In the year that the budget is spent, funds are those approved during the previous fiscal year.

Model of Utility Finance

For local water organizations the general approach to financing and funding follows a similar model for water, wastewater, and stormwater. Figure 19.1 illustrates the capital and operational parts of this approach. It was developed by the director of the Fort Collins water utilities to explain their financing model.

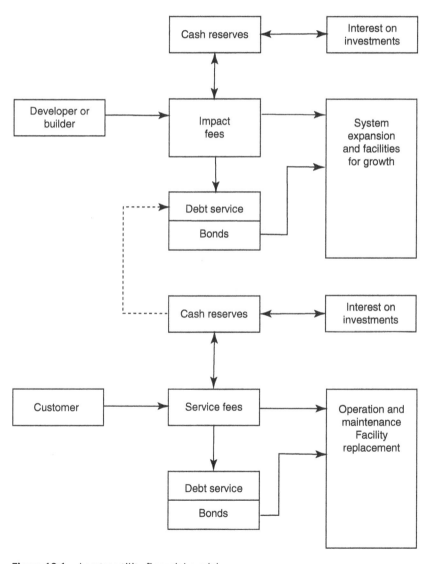

Figure 19.1 A water utility financial model.

The top part of the diagram shows how a developer or builder pays plant investment fees to finance system expansion. The bottom part shows how customer service fees pay for operations and maintenance. The diagram shows a distinction between system expansion and system renewal. Expansion is financed by developers and builders through a "growth pays its own way" approach financed by the plant investment fees. System renewal or facility replacement is financed by customer service fees. This is appropriate in that it requires those using the system to pay for its replacement while new connections pay for the initial expansion. Cash reserves and debt service payments are also shown on the diagram.

Rate-Setting

Setting customer service fees for the actual cost of water service involves rate-setting, which is a common and necessary task in public utilities. Adequate fees are needed to ensure that the utility receives enough revenue to cover the capital and operating costs of source of water supply, treatment, storage, and distribution, as well as associated wastewater and stormwater systems. Rate-setting methods for water supply, wastewater, and stormwater are somewhat different, but follow many of the same principles.

Utilities also need to fund their development of organizational capacity, which involves technical capacity for its infrastructure development and operational responsibilities, managerial capacity, and financial capacity that is adequate to acquire and manage enough funding for system development and operations.

Principles of utility rate design have evolved and are similar for water and electric power utilities. As public water services began prior to public electric services, some of the principles stem from their rate decisions. In any case, rates should be simple, understandable, and acceptable to the customers. They should provide an adequate and stable revenue stream. They should be fair in assigning costs of service, which creates equity among customers. Rates should also promote efficiency in use of water resources and infrastructure.

The process of rate setting is to establish revenue requirements, allocate them to customer classes, and design the rate structure to provide the required revenues. Establishing revenue requirements means to assess costs to provide the services. Customer classes are groups of similar water users who should be charged using common rate schedules, and they can be residential, commercial, and industrial customers, as in the case of water supply.

There are different ways to design the rate structure, and the most appropriate one in a given situation depends on the age of the system, complexity of treatment, service area size, mixture of customer categories, climate, geography, and water demand, among others. For example, a system serving a small number of residential customers will require a different rate structure than a larger system involving a mixture of types of customers and in the different geographical setting.

The principle of cost-of-service pricing is well-established in rate setting. It means to charge at the marginal cost of supplying the service. The problem with it in setting water rates is that it does not penalize high water users and incentivize conservation.

Types of rate structures include flat rate, uniform rate, decreasing block rate, and increasing block rate. Other types of rate structures involving water budgets are also used. Examples of rate-setting using these types of rate structures are given in the problem set at the end of the chapter.

Connection Charges for System Expansion

Water connection charges are used to fund costs of capacity investments such that "growth pays its own way." They are mainly used where regions confront population growth, water scarcity, and resource vulnerability. They help manage future water demand and development patterns. Related terms are growth fees, tap fees, capacity charges, system development fees, impact fees, or plant investment fees. The same types of charges may be used for other infrastructures, like roads, parks, and sewer systems, among others.

The concept is that new customers generate requirements for new infrastructure which should be financed through their payments. Utilities can also use connection charges along with rate structures to incentivize water conservation. The fees can become quite large, particularly in areas requiring scarce water rights and this can add large sums to the cost of housing.

To calculate the required fee levels, the cost of the system expansion must be estimated and spread over the new customers. It will not be feasible to calculate the added cost for each new customer, so it will be necessary to make projections over several years and with addition of multiple new customers to develop average fees.

Connection charges usually consider water use as predicted by lot size and building characteristics, although some utilities may base connection charges only on meter size.

In the Fort Collins financial model, plant investment fees are shown to finance system expansion. Customer service charges are used to finance system renewal. This approach places costs of added infrastructure on new users of the system and places responsibility to sustain the system on current users.

Debt Financing

Use of debt financing through bonds and loans is the most common way to finance infrastructure development. Use of loans to finance water systems is normally from government financing authorities such as infrastructure banks. These are favored for water supply and wastewater as will be explained in the examples later. Funds are normally limited and might not be adequate for larger projects.

In bond financing, the use of revenue bonds to finance capital investments is usually favored over general obligation bonds. The payback of revenue bonds is from income of the enterprise, which is a businesslike approach in that it supports cost recovery and is self-supporting. General obligation bonds may be repaid from revenues, but they are backed by the general tax base. In bonds, the required debt is issued by underwriting firms and periodic interest payments flow back to the bond holders. The bond market provides a mechanism for investors to trade their bonds. The utility can refinance bond debt to take advantage of lower interest rates when they occur.

Financial Assistance Programs for Utilities

It is expected that some water authorities will need financial assistance through grant programs if they are to afford necessary expansion and improvements. As would be expected, these are sought more by systems facing large expenditures but have inadequate revenues. Although the federal government provided a large-scale grant program to implement wastewater treatment after the Clean Water Act of 1972, such grant programs no longer exist. However, small and special purpose grant programs may be available in selected cases.

There is no national clearinghouse to identify grants for water systems, which are local in character and site-specific. This is different from highway grants, which generally flow to state agencies. For the most part, grants have given way to loans are the preferred method of capital financing. In some cases, small grants may be bundled with the loans made available to disadvantaged communities. The only way to find such grants is to search for them on a case-by-case basis.

For drinking water and wastewater systems, USEPA is a good source of information, which can be accessed by a simple Internet search or by a search of the agency's website. Another potential source is the USDA's Rural Utilities Service. The Department of Commerce may have grants through its Economic Development Administration, and the Department of Housing and Urban Development has a Community Development Block Grants Program that might offer grants to partially fund water systems.

Larger states, particularly California, have some grant programs. A good source of information is the state Department of Water Resources. Most states have a central office to coordinate financial assistance that flows from federal programs, as well as any state-financed loan or grant programs. These can be called loan authorities, financial authorities, development banks, infrastructure banks, or similar titles. Normally, staff in these authorities know where to look for financial assistance.

Accounting and Financial Analysis

Financial accounting and analysis are needed when water managers confront financing issues related to their projects and operations. Managers require management accounting, which provides the information they need to make decisions and control their organizations. Few water managers study accounting, and they need some background in how it works for public sector issues. Although, water organizations mainly use government accounting, their accounting tasks are similar to those of other organizations, requiring analysis of labor, materials, and other expenses.

Financial accounting has a logical structure to track and record the "stocks and flows" of money. Stocks are inventories of any quantity, such as money, fuel, water, or other commodity. Flows are rates of change. In basic accounting, transactions are recorded through bookkeeping, which enables preparation of financial statements and reports, which support decisions of managers, directors, customers, and regulatory agencies.

At the basic level, accounting information includes such records as time sheets, vehicle logs, receipts, and other reports of daily operations. Results are aggregated to financial reports that are used in decisions about financial control.

Managers normally see financial documents with budgets, which are financial plans. Then, financial statements report on results after a plan is implemented. The two basic financial statements are the income statement. Income statements report the differences between revenues and expenditures over a period such as a month or a year. Information about assets and liabilities is reported on the balance sheet. Whereas an income statement is like a report of water flow and additions or deductions from a reservoir, the balance sheet is like a snapshot of how much water is in the reservoir at the end of the year, along with how much is owed to users and how much is expected from others at that point in time.

The annual report is useful to report financial results and to summarize the accomplishments of an organization. They focus management's attention on results achieved and financial health. In businesses, the primary goal is the "bottom line," and the profit of the business's operations. In government agencies or enterprises, the approach is to present results more objectively without a focus on financial profit but on profit to the citizenry as a whole. This is what Triple Bottom Line reporting is about.

Fund accounting is used by public sector organizations to show accountability rather than profit and loss. Funds are self-balancing accounts to report expenditures by designated purposes. Examples are the general fund (like in a city), which accounts for municipality operations. Enterprise funds are used for separate services to external parties, such as utility services. Capital project funds are used to track projects from a variety of revenue sources, and debt service funds collect fund levels for repayment.

Infrastructure systems are capital intensive, but accounting for fixed assets has been neglected in management accounting. Fixed assets are depreciated by accountants, but depreciation relates to tax obligations more than it does to condition of assets. Ideally, data on depreciation could be used to inform the need for system renewal.

Financing Water Systems, Services, and Projects

Each of the water systems and service areas has a unique approach to financing. Water supply is the oldest local public utility and many of its financing and funding principles also apply to wastewater and stormwater. Other water management categories, like irrigation or hydropower, have distinctly different issues to contend with.

Joint Projects

Financing joint multipurpose projects requires that you assess costs in a fair way in proportion to how different parties benefit from a project or a service. Most projects have costs that are necessary for all parties (joint costs) and costs that are identifiable with beneficiaries (separable costs). In multipurpose water resources projects, the "separable costs-remaining benefits method" has been developed to separate joint and separable costs and allocate separable costs per benefits received.

Take, for example, a Corps of Engineers multipurpose reservoir with three purposes: water supply, flood control, and hydropower. The hydropower might be produced by the reservoir and sold to utilities on a wholesale basis so that separable costs are financed by user fees. The flood control might be jointly financed between the federal and state governments, with appropriate allocation by negotiation. The water supplies might be financed through sale of the water to local governments through long-term contracts. The allocation of costs will be done through negotiation. Another example could be a drainage and flood control project for part of an urban area. Some benefits will accrue to land developers and some to the public at large. The city negotiate with the developers per its policies and goals. Still another case could be allocation of costs by customer classes, say for a water and wastewater utility. This will involve application of rate studies and principles among the services and zones of the city.

Water Supply

Water supply utilities evolved from early roots as private businesses to sophisticated public organizations. Some began as cooperatives or departments of local governments and adopted similar principles. Water managers instituted fiscal measures to operate their systems on a sound basis and the lessons were embodied in AWWA's first manual, M1 Principles of Water Rates, Fees and Charges, which was first published in 1954.

Water utilities provide potable water supply to residential, commercial, and industrial customers, and they provide water for public uses, especially fire protection. They are capital-intensive, and about 70% of their assets comprise underground distribution systems. Sustaining financial sources to support their infrastructure and organizations is a continuing priority for water utilities.

The utility operating budget will include costs for personnel and support items such as building maintenance, vehicle operations, and supplies. Costs will be allocated to source of supply, treatment, and distribution, as well as to associated services such as a laboratory. Water usage can be measured, so users can be charged according to use. This facilitates the use of water rates and charges to fund operations.

Prior to widespread adoption of water meters, most systems charged a flat rate, which might differ by connection size or property type. With water metering, marginal cost pricing could be used for decreasing block rates. This approach does not promote efficiency in water use and can encourage waste of water, so increasing block rates have been introduced to provide incentives to conserve water.

Figure 19.2 shows three rate concepts, decreasing block rate, flat rate, and increasing block rate. The word block is used to enable a constant charge across a range of water use.

This is an example of a monthly charge using an inclining block rate for a three-person home in the City of Fort Collins. The use is high due to outdoor watering

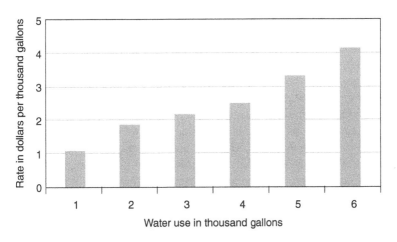

Figure 19.2 Water use rate structures.

in a dry month and per-capita use was 236 gpcd. This will drop to about 100 gpcd in winter months, and the average for the year will be near the city's norm of about 140 gpcd.

		Charge per 1000 gal, $	Usage, gal	Charge, $
Base charge				18.30
Tier 1	0–7000	2.83	7000	19.81
Tier 2	7001–13000	3.26	6000	19.56
Tier 3	Over 13000	3.75	8980	33.68
			21980	91.35

Utilities also have capital budgets, which show plans and decisions about investment for infrastructure and equipment. Capital needs may be generated by need for expansion, obsolescence, deterioration, regulatory change, or land use changes.

A general picture of drinking water infrastructure needs is available from a survey-based assessment of 20-year investment needs for replacement and rehabilitation conducted by USEPA. The federal government has never had a substantial program to subsidize funding needs of water supply systems because these are viewed as a local responsibility. However, the federal government decided in the 1996 Amendments to the Safe Drinking Water Act to create the Drinking Water State Revolving Fund and the assessment is used to support federal appropriations for the Fund.

Funds from the Drinking Water State Revolving Fund are distributed to state governments, which administer revolving loan programs. When funds are available, utilities can borrow from these state revolving funds, although more capital funding is from other debt, especially revenue bonds. Plant investment fees can also be a significant source of capital funding, as can payment from reserves or pay-as-you-go (from income) to create sinking funds for investment. For small systems, some grants may be available.

Most utilities provide effective services, but they face challenges to pay for aging infrastructure, including in shrinking cities, and the need to finance capital issues such as replacement of lead service lines. While some utilities have impressive capacity and resources, many others are small and lack capacity to confront the issues they face. While needs are increasing, water conservation programs may reduce revenues and require rate increases to sustain adequate funding for system operation and maintenance. For these reasons, innovations in rate structures are being introduced to sustain revenues, while addressing challenging issues.

Water utilities can generate substantial revenue, and cities usually charge a "payment in lieu of taxes" (PILOT) to offset the tax-exempt advantage of a public utility over a private utility. Also, if city staff perform services like customer billing, they may levy an administrative charge to tap into the revenue stream. These mechanisms may enable the city to avoid tax increases and can be politically attractive. It is important that they are not abused to the extent that the water utility's revenues are drained and funds are lacking for capital renewal.

Wastewater Management

Today's wastewater utilities evolved on a different trajectory than water supply utilities. Beginning with the advent of indoor plumbing, cities needed wastewater collection systems to transport sewage away from buildings, but wastewater treatment plants were rare. Beginning in the 1950s, with construction of more wastewater plants, the management responsibilities of local governments increased. Whereas collection systems mostly need repair crews, wastewater treatment added more complexity.

The Clean Water Act of 1972 authorized subsidies to construct wastewater treatment plants and led to new requirements for operations and maintenance. Cities initiated wastewater utilities and developed formal rate structures. When wastewater systems were only collection sewers, their maintenance could usually be handled through tax-supported city budgets. With more complexity, more complex rate structures were needed and USEPA required an approved rate structure prior to awarding grants for wastewater treatment subsidies.

Wastewater rate structures are different than for water supply and they vary among localities. Utilities should have cost recovery plans that address impacts of different user classes on the system, including residential, commercial, and industrial classes. The theory is that wastewater customers can be charged by some metric related to water use, such as winter quarter average use, when no outdoor use is happening. Commercial and industrial customers may also be charged by water use, but some may face surcharges on the basis of the strength of their wastewater.

As an example, a wastewater charge for the city of Fort Collins can be illustrated. Residential wastewater rates include variable charges based on winter quarter average water use and fixed charges for a connection to the collection system even when not in use. Most wastewater costs are fixed to cover operation and maintenance of collection systems and wastewater plants. Costs are calculated annually and applied to all accounts. In 2020, the residential base charge was $18.86 per month, and the volume charge per 1000 gal (3785 l) was $3.66. Thus, if the winter quarter average water use for a three-person home was 100 gpcd (380 l) for a 30-day month, the volume charge would be $32.94 and the total monthly charge would be $51.80.

Capital financing for wastewater systems is similar to that for water supply. USEPA prepares a periodic Clean Watersheds Needs Survey, which informs

Congress about funds needed for the Clean Water State Revolving Fund. This fund was created by the 1987 amendments to the Clean Water Act to replace the USEPA Construction Grants program that subsidized treatment plants.

Wastewater utilities face similar issues to water supply utilities. They must meet standards while dealing with important pollution issues, including system upsets, sanitary sewer overflows, and combined sewer overflows. They must do this while managing complex and aging infrastructure with funding challenges and a continuing need to strengthen its workforce. Many systems are small without capacity to confront the technical and financial issues they face. One unique feature of wastewater utilities is their attractiveness to become "utilities of the future," which would incorporate multiple waste management streams including food waste recycling.

Another issue encountered in some wastewater systems is implementation of water recycling. Due to water shortages, some utilities have implemented water reuse systems to treat wastewater and make it available for either potable or nonpotable uses. Potable systems are rare but are needed in some cases. Financing water reuse systems takes the same approach as water supply, but it can be difficult to establish rates because demand for the water is more problematic. In some cases, utilities must provide incentives to customers to use nonpotable water. Another problem in reuse systems is the need for additional infrastructure, which can be difficult to finance.

Stormwater

The need to finance stormwater systems evolved as their complexity increased (see Chapter 4). Stormwater systems provide drainage and water quality services and calculation of charges must consider both. These calculations differ from rates for water supply and wastewater, although principles of rate design are similar.

Stormwater systems involve primarily conveyance and additional infrastructure for water quality control. The conveyance systems provide for minor drainage and major drainage, which is like flood control in urban channels. Revenue requirements to operate them focus on the need for maintenance, along with monitoring and reporting. Capital financing is mostly needed to construct, expand, and modernize systems.

Current approaches to stormwater finance started in the 1970s when a taxpayer revolt in California forced local governments to use fees instead of taxes for services. By the 1980s, this movement converged with required water quality controls on stormwater discharges and created conditions to enable stormwater utilities. By the 1990s, it was clear that the stormwater utility movement would grow, and today it is well-established, although not without controversy. The essence of the controversy is whether stormwater fees are taxes in disguise or if they are bona fide service fees based on benefits to users. This question has been litigated a number of times, but without a universal outcome. In some localities, stormwater utilities remain criticized as using "rain taxes."

The theory of stormwater charges is that conveying runoff from land provides a benefit and a basis to estimate a cost of service. In the case of water quality, the assumed benefit is similar to wastewater in that the pollution caused is managed through the stormwater service. The theory generates obvious questions in that large parts of cities produce drainage and pollution that are not connected to individual properties. This means that general benefits usually financed by taxation are included in the charges, as well as special benefits that accrue to individual properties.

Despite continuing controversy, many stormwater utilities operate on a continuing basis. To illustrate an example from Fort Collins, rate factors are computed from the cost of service to operate the stormwater utility. The monthly 2022 rate for single-family lots was $0.004\,53 \times$ runoff factor $\times$ square feet up to $12\,000$. Over $12\,000\,\text{feet}^2$ ($1115\,\text{m}^2$), the rate was $0.001\,13 \times$ runoff factor $\times$ square feet. A runoff factor of 0.4 for light residential development is based on a percentage of impervious area that is judged to be typically 31–50%. This results in a monthly charge for a $10\,000\,\text{foot}^2$ lot of $0.4 \times \$45.30 = \$18.12$.

The rate factors used by Fort Collins are shown in this table:

Rate factor	Percent of impervious area	Rate factor category based on land use
0.25	0–0.30	Very light
0.4	0.31–0.50	Light
0.6	0.51–0.70	Moderate
0.8	0.71–0.90	Heavy
0.95	0.91–1.0	Very heavy

Stormwater programs face political controversies that are different from water and wastewater. The allegation of the "rain tax" plagues some localities, which might also face affordability issues. Continuing pressure to improve stormwater quality by installing best management practices is another challenge, along with changing land uses and climate change. Lacking treatment facilities, aging stormwater infrastructure does not create a problem as severe as water and wastewater.

Utilities and customers often clash over charges, and methods of calculation and rates vary. Inaccurate or outdated data leads to problems and legal challenges. Utilities need data on building footprints, driveways, and other surfaces to estimate impervious cover. The data can be collected by aerial photography or satellite imagery, and it should be validated via fieldwork. It is critical to define impervious and pervious surfaces with clarity, and utilities should acknowledge the impervious areas not benefitted by the systems.

Financing the operation of combined sewer systems is a special case involving combined stormwater and wastewater benefits. In most localities, there seems to be little controversy over how to allocate costs to wastewater or stormwater services, but controversies can arise.

Irrigation Systems

Financing and funding irrigation systems are driven by system types. On the local basis, irrigators are likely to belong to a mutual irrigation company or to provide their water by self-supply. Mutual irrigation companies levy user charges on the basis of cost recovery, but the costs will normally be limited to operation and maintenance when systems are old. Self-supply systems are likely based on wells and center pivot sprinklers or other technologies.

In the western United States, irrigation operations of the Bureau of Reclamation provide other cases where water districts may operate infrastructure under contract with the federal government and pass their charges on to users accordingly. When water districts combine irrigation with other services, especially electric power, funding streams become more complex. For example, electric power users may subsidize irrigators, which may not generate controversy if the users are the same.

In lower income countries local irrigation systems may be operated by water user associations which function like mutual irrigation companies. In many cases, these must operate with little organization or funding and are sometimes operated by central or regional governments.

Irrigators face increased competition for water and are challenged to obtain adequate revenues by water pricing. The connection of irrigation to farm and food policy introduces additional complexities.

Flood Risk Management

The financing of flood control began with an early focus on building dams and has now shifted to flood insurance (see Chapter 21). This means that, while financing some upgrades and renewal of flood control infrastructure, financing the flood insurance program is a major priority. At the federal level, the USACE maintains a number of flood control reservoirs and levees. Some are owned by state and local governments and water management districts as well. Most flood control works are owned by local governments and how they are financed depends on their organization.

For example, the Miami Conservancy District was formed after a massive 2013 flood disaster. The State of Ohio passed a Conservancy Act to establish the District, which now owns and operates dry flood control dams and retarding basins. Financing is via assessments on properties and jurisdictions who benefit from District programs and facilities. Another example is the Mile High Flood District in Denver. It has a budget of about $65 million, and is funded by county property tax assessments.

Instream Flow Management

Financing water management for instream flows involves navigation, hydropower, water quality, and fish and wildlife. Recreation may also be involved, and channels are needed for conveyance of water from place to place. These different categories of water management do not have a standard financing approach but will involve a mixture of methods.

Water management for navigation is handled mainly by the USACE and involves a navigation trust fund with some repayment from user fees. Hydropower is managed both by private electric companies and the government and water management is bundled with other system expenses. In case of the private systems, the Federal Energy Regulatory Commission imposes requirements that affect costs. The costs for instream water quality and fish and wildlife are embedded in other system costs. For example, monitoring for water quality will be conducted by utilities and regulatory agencies. Water for fish and wildlife is normally imposed through regulation, but in some cases, and stream flow rights may be purchased.

Water Management Agencies

The government agencies involved in water management are diverse in nature and have different funding streams. An example is the USGS budget at the federal level, which includes a mixture of Congressional appropriations, fees for services provided, and cooperative cost-sharing revenues. Other agencies, like the USACE, are primarily funded through Congressional appropriations.

State agencies, like the regulatory programs in every state, are funded through a combination of federal grants and state appropriations. The federal grants provide a baseline and state appropriations will normally make up the difference such that the programs have sufficient revenue to pay staff and conduct operations. The few state agencies that actually own and operate water management facilities, like the California Department of water resources, and of budgets with multiple funding streams often including user fees.

Private Water Companies

Private water companies have a long and important history in the United States and they date back to the first organized US water utilities, beginning around 1800. There are many local stories of US private water companies operating in growing cities as the United States emerged from its status as a rural colony to become an urbanized and independent nation. While many of these early private water companies went public as cities grew and local politics became more sophisticated, the formation of private water companies continued. On the international front, some private water companies have grown to giant scales. While there is an ongoing debate over the relative merits of public versus private ownership of water systems, it seems clear that privately owned water companies can deliver safe water competitively and operate as responsibly as publicly owned water companies. Whether they can be more efficient and hold costs down would depend on a case-by-case basis. The operation of private water companies focuses on the utility nature of water and wastewater services, and it is hard to see how these for-profit organizations could go very far toward responding to environmental or social needs and still serve their investors well. They might be compelled by their regulators to take certain actions, such as maintain a program for rate relief of low income residents, but these would not be very consistent.

Private water companies are represented in the National Association of Water Companies (NAWC) (2010), which is their trade association. It reports statistics from USEPA that some 73 million Americans get their drinking water from privately owned companies or a municipal utility operating under a public–private partnership. These partnerships would normally involve operating contracts. The number served by utilities that are entirely investor-owned or in other private ownership would be fewer. NAWC reported further that private water companies have about 100 000 miles of distribution main, which is between 5 and 10% of all mileage in the United States. If it is 10%, this suggests that private water companies serve around 30 million customers, based on a total US piping of 1 million miles. The quantity of piping is discussed in more detail in Chapter 5. Further, the private water business is reported to be about a $4.3 billion per year business, which suggests, based on US averages, that it supplies on the order of 30–35 million people.

Among water-handling organizations, the private water companies are somewhat unique in being the only ones to be regulated by state public utility commissions (PUCs). PUC regulation of water companies is similar to that imposed on electric power and natural gas utilities, along with other monopoly privately owned public services. While PUC regulation is imposed on private water companies, publicly owned water utilities are regulated by local governments based on political choice, although they are regulated by the USEPA for health and environmental goals. For rates, local elected leaders take public interest into account when making rate decisions and try to balance the views of businesses, individuals, and nongovernmental organizations, as well as the needs to balance health and the environment and to sustain the economy.

Affordability of Water Services for Individuals and Communities

Affordability of water services is a major issue for communities and individuals, along with their concerns about housing, food, water, and energy affordability. For individuals, it means the ability to pay your water bill without compromising your ability to pay for other goods and services you need, and for communities it means the cost of water facilities compared to statistics like average income of residents and other financial ratios.

There is a built-in conflict for water utilities, who must sustain high quality services while addressing the needs of low-income customers. Sometimes lifeline rates are used to provide a subsidy through discounted rates for low-income families, seniors, and disabled persons.

The cost of living affects communities and individuals, who may be under pressure even while communities as a whole are not. The US Environmental Protection Agency (EPA) has published criteria for affordability of water supply services, but it addresses community affordability and the situation for individual customers is different. EPA's criteria for average conditions in a community are based on a comparison of average annual cost of household water services to the median household income.

EPA has not adopted a measure of how much a household can pay for water services before they are unaffordable. EPA initially set a 2% of median household income affordability level for water supply by comparing its cost to expenditures such as alcohol, tobacco, and energy and subsequently increased the level to 2.5%.

Within the water community, there is consensus that utilities should operate as businesses, but they will also be confronted by affordability issues for low-income customers.

Drinking water utilities have recognized the problem that individuals have and addresses it through a policy statement on "Discontinuance of Water Service for Nonpayment." AWWA opposes cross-subsidy but recognizes that certain circumstances may require some flexibility.

Although affordability definitions did not originally address individual low-income customers, utilities took it on themselves to address them with customer assistance programs (CAPs). Income distribution can vary widely across different districts and hardship can be concentrated in a few neighborhoods. Addressing social issues of this kind is important to create healthy communities, but assistance programs for water bills can be a challenge because some many customers live in single- or multi-family rental buildings or public housing and pay for water through rent or a home maintenance fee. Often, they live in buildings that are out-of-reach for conventional utility assistance programs. They are likely to be in poverty, have a disability, speak English as a second language, and have lower-than-average education levels.

Utilities are not expected to become social services agencies but customer assistance should be integrated into business processes. Categories of customer assistance programs are: Bill discounts; flexible terms; lifeline rates, temporary assistance; and financial assistance for water efficiency. Public assistance funds are a good strategy, with emphasis on partnerships to address multiple needs of low-income residents, including those not billed directly by utilities.

State statutes range from prohibiting assistance from rates to mandating assistance programs. Also, public- and investor-owned utilities may be treated differently.

Ultimately, the need for fair treatment of low-income people and community improvement is a global issue, and the water sector can contribute. By implementing definite strategies within their business processes, utilities can provide leadership in addressing important and shared issues in their communities.

Questions

1 Explain situations where a water manager must use knowledge about budgeting and accounting for funds and when the manager would be involved in financial arrangements for capital expenses.

2 What causes financial problems in many small water systems? What impacts do these have on service levels?

3 Among the financing mechanisms of taxes, tariffs, and transfers, which is most appropriate for water supply utilities?

4 In the Fort Collins utility financial model, explain how operations, system expansion, and infrastructure renewal are funded.

5 Search the Internet to find the budget for a US national water agency, like USGS, for example, and summarize the main sources of funds and types of expenditures.

6 Explain how the Planning-Programming-Budgeting System (PPBS) works in general management of water utilities and in capital improvement planning specifically.

7 Are rate-setting methods for water supply, wastewater, and stormwater the same? Explain any differences.

8 In rate-setting, what is a customer class? Give an example. Explain the concept of cost-of-service pricing in rate setting.

9 For a local water supply system, give examples of where rates, fees, taxes, transfers, and debt might be used.

10 Explain how conservation pricing works and why using cost-of-service to set urban water rates is not always appropriate.

11 Why can water supply services be supported by user charges more readily than wastewater services?

12 Explain the concepts of flat and block rates for water use.

13 What is the most logical way to finance water supply system renewal?

14 Why have water connection fees become so expensive in some places and what is the impact on the cost of housing?

15 Explain the basis of cost-of-service in rate-setting for stormwater utilities.

16 What is meant by "equity" in the design of water rates?

17 How do "payment in lieu of taxes" and administrative charges affect funds available to manage a water supply utility?

18 If water services are subsidized from taxes in a progressive tax system, how does this shift costs from water users to people according to the income?

19 What is the policy basis for connection charges levied on urban developments?

20 Explain situations where financial assistance using grants might be available to water utilities and what the sources of funds might be.

21 How do water financing authorities work at the state level?

22 What are meant by "pay-as-you-go" and "pay-as-you-use" capital financing?

23 Is debt financing "pay-as-you-go" or "pay-as-you-use"? Explain your answer and any difference this makes in who pays and who benefits.

24 Why is debt used more often to finance capital expansion in water utilities rather than through savings and reserves?

25 Give examples of O&M and capital costs in a water utility.

26 Explain the financing problem of replacing lead service lines and why this was included in the infrastructure bill of the Biden administration.

27 What difficulty might arise in charging fees for recycled water?

28 What are the main needs for funding for the management of wastewater collection systems?

29 Why is development of user charges for wastewater more complex than for drinking water? How, in your opinion, should it be done for residential, commercial, and industrial customers.

30 Why were subsidies needed to create a fast start to wastewater plant construction after the Clean Water Act was passed?

31 Compare loan versus bond financing in terms of funds available to the borrower.

32 What is the World Bank and how is it involved in financing water systems?

33 What is a revolving fund? How does it work? Where does it get its base funding?

34 Compare the operation of USEPA's CWSRF program to the way a development bank works.

35 Do water and wastewater state revolving funds provide subsidies? Why or why not?

36 What was the driving political force behind creation of stormwater utilities?

37 Where would the funding normally come from for a small flood control project in a city?

38 What is the major funding source for the National Flood Insurance Program? What are the major payouts for?

39 Why was the National Flood Insurance Program established?

40 What is the major source of funding for USACE flood control projects?

41 What is the source of revenue for mutual irrigation companies?

42 Why is it difficult to charge irrigators the full cost of water?

43 Where do the funds come from to support capital needs of Bureau of Reclamation projects?

44 Does management of stream fisheries involve major costs to water agencies? Explain your answer.

45 Other than wastewater treatment, what costs occur for management of water quality in streams?

46 Where do the funds to manage hydropower facilities come from?

47 What are the revenue requirements for navigation water management? Who manages the funding for navigation water management?

48 Explain the purpose of affordability programs for urban water services. How are organizational and household affordability metrics different?

49 What is an enterprise fund in the context of urban water supply?

50 In the water supply field, what is an example of a government enterprise?

51 Is privatization appropriate for water supply services? Wastewater? Stormwater? Explain.

52 What is the nature of the rain tax controversy in stormwater funding?

53 What are the main differences in the ways that private water companies and government-owned utilities manage their finances?

Problems

Water Rates

A family of four people uses an average of 200 l of water per person per day. Water charges are computed as a base charge of $20; plus $3.00 per thousand gallons (TG) from 0 to 3000 gal; plus $4.00 per TG from 3000 to 6000 gal; plus $5.00 per TG for all usage above 6000 gal/month. Compute the charge for the family for a 30-day month.

Wastewater Charges

A home with three residents has water uses during the winter quarter (January, February, and March) of 8000, 9000, and 10 000 gal. Wastewater rates are computed as a base charge and variable charge that depends on the winter quarter average. If the base charge is $18 and the variable charge is $4 per thousand gallons, what will be the monthly charge?

Stormwater Rates

Stormwater charges are based on impervious area. For a 10 000 square foot lot, compute the monthly charge for the following conditions:

Monthly Rate = $0.00453 × runoff factor × square feet up to 12 000
Runoff factor for single-family lots = 0.4

Plant Investment (Growth) Fee

The infrastructure to supply drinking water in a medium sized city costs $10 000 per capita to develop. What would be the plant investment fee levied on a duplex if the expected average occupancy rate is 2.25 persons per unit?

Water Charge Calculation

A city of 50 000 people has water infrastructure that would cost $10 000 per capita in current dollars if it had to be totally replaced. The city decides to renew it partially with an investment of $100 million to be implemented immediately. Using the data below, compute the water charge in dollars per thousand gallons that will be required to finance the system.

- Annual operations and maintenance cost: $100 per capita
- Average water use: 150 gpcd
- Planning period: 40 years (no other investment required in this period)
- Interest rate to borrow money: 4%.

If you need to make any assumptions, explain them.

Infrastructure Bank

An infrastructure bank will loan $5 million to a water utility at a subsidized rate of 3%. The infrastructure bank must raise its funds by selling tax exempt bonds on the open market at 4%. The term of the loan to the utility is five years. How much money must the bank spend over the five years to cover this subsidy?

Capitalize Maintenance

A wastewater department wants to capitalize annual maintenance fees of $5 million. It applies to the central government for a grant to fund this capitalization, at 5%. What must be the size of the grant to do this?

Comprehensive Financial Analysis

Conduct a financial analysis of water system operation for the hypothetical City of Smallville over the next 10 years. Use the given data or make assumptions as needed. Analyze O&M, capital costs, and cash flows. Prepare a:

- Schedule for capital improvements
- Spreadsheet showing your financial analysis and cash flows
- Pro forma income statement (projected statement) for year 10
- Pro forma balance sheet for 31 December of year 10
- Graph of how system value varies over the 10-year period

The situation is that Smallville's water system currently serves 100 000 people and is to be expanded to handle a population influx of 5% per year for the next 10 years, starting now. Land use is mixed residential and commercial, no industry. Base your estimates on residential demands and assume that commercial water use adds 15% to residential use. The system is 15 years old and has depreciated up to now on a 30-year depreciation cycle at 3.33% per year on a straight-line basis. New and renewed system components will continue to depreciate at the same rate. You plan to expand the system to accommodate growth and simultaneously maintain and renew the existing system during the period. Your system expansion will be staged so that half is built now and half in five years. When you also invest in system renewal to overcome depreciation, the investments are added to costs of system expansion. System renewal scheduling is governed by the rule that current system value must not fall below 50% of replacement value.

You will recommend how to finance the expansion and renewal with funding from plant investment fees, water use fees, sales tax revenues, and property tax revenues. Take a loan for the first part of the construction at year 0. Issue bonds for the next increment, starting at year 6 (in five years). You may vary this if you choose different capital financing vehicles. Loans are "revolving loans" and come from an infrastructure bank. Annual loan payments begin end year one and continue for 10 years. Commercial property has 25% of assessed valuation of residential property.

Data for the exercise are given in this table:

Current population	100 000 (33 333 households)
Rate of population growth	5% per year for 10 years; 0% after that
Per capita water usage (average)	150 gpcd
Land use	Mixed residential and commercial
Planning horizon for capital improvements	10 years to meet demands; 30 years for system life
Plant investment fee	$5000 per house connection
Current water fees	$2.50/1000 gal
Property tax dedicated to water system improvements	0.8 mills
Assessed valuation residential (market value ×0.2)	$980 million
Sales tax dedicated to water system	0.8%
Current anticipated taxable sales	$800 million per year
Interest rate on loan (due in 10 years)	8%
Interest rate on bonds (use 20 year life)	6%
Projected inflation rate	0%
Capital cost of new or replacement system	$3000 for each new person
Current value of existing system (average age 15 years)	Replacement value less 15 years depreciation
Capital improvement goal	System value not below 50% of replacement
Depreciation of assets	3.33% per year
O&M cost	$50 per capita per year

Rate-Setting

This exercise illustrates how to estimate capital and operating costs of water, wastewater, and electricity systems, allocate costs, and set user charges (rates) for the services while satisfying efficiency and equity requirements. A utility organization offers three services to city residents: municipal and industrial water supply, wastewater services, and electric power. It also sells excess electric power to an outside rural utility, and it leases its surplus water to irrigators. Thus, five services generate revenue: water supply, wastewater, electric power to the city, wholesale electric power to a rural utility, and leased water to irrigators.

Your tasks are to estimate the capital and operating costs of the utility and allocate them to classes of the utility's rate-payers. Capital and operating costs and related data for each service are given below. These are your tasks:

- Compute the total annual costs of the separate services.
- Show how to allocate charges to cover the costs without subsidies, that is, for the utility to be self-supporting.
- Submit a memorandum to explain your results with supporting data in some detail. Include in the memorandum an explanation of how you ensure efficiency and equity to all classes of users.

Customer Rate Classes

- Urban water users, population 100 000 (assume an average year and all residential users at 175 gpcd and no commercial and industrial users)
- Urban wastewater, population 100 000–100 gpcd in wastewater generation
- Retail electricity customers are the urban population of 100 000 (they use all electric power after sales to the wholesale rural electric utility)

External Customers with Prices Fixed by Contracts

- A wholesale electricity customer, a rural electric association, buys 500 million KWH per year. You charge them at 50% of your retail electricity rate per KWH to utility customers.
- Irrigation water customers have a contract to get 50 000 acre-feet per year and pay $100/AF. This fixed cost to the irrigators and revenue to the utility is for an "interruptible supply" that averages 50 000 AF/year. If the supply is interrupted, the irrigators get six months' advance notice. If more than 50 000 AF are available, they can purchase this water at $25/AF. You do not know the amount available for any year but you assume that an average of 100 000 AF/year are available and that the urban users are at an average level of 175 gpcd.

Cost Data

- *Raw water rights purchase cost to supply both irrigation and urban water.* The utility owns water rights of 100 000 acre-feet for urban supply and irrigation supply. The purchase (capital) cost is $3000 per acre-foot and operating cost is $25 per acre-foot per year for the raw water.
- *Urban water supply infrastructure and operating costs.* Capital cost $1250 per capita; Operating cost $50 per capita per year (not including raw water)
- *Urban wastewater service infrastructure and operating costs.* Capital cost $1500 per capita; Operating cost $75 per capita per year.

- *Electric power.* Total generation is 1.2 billion KWH per year (of which 500 million goes to the wholesaler and the remaining 700 million is for the utility's retail customers). Capital cost is $2000 per installed kilowatt of capacity and the O&M cost is $0.03/KW-HR. (Note, you need more capacity than the average consumption to meet peak needs. Use a reserve factor of 1.3 to multiply by the average KW demand and use this to compute the capital cost).
- *Interest rate and project lifetimes.* For capital is 4.0%; use lifetime of 30 years for any economic studies of the cost of infrastructure.

Cost Allocation to Utility Customers

- Basic water supply for urban customers, rates per 1000 gal units.
- Wastewater charges per 1000 gal of wastewater generated by urban customers.
- Electricity charges, per KWH supplied. Assume that retail consumption = about 0.8 KW/capita on a steady basis. This may vary from year-to-year and is an average only. (Recall above that you must have a generation capacity of 130% of the average use.)

Cost Allocations to External Customers

- Electricity charges to wholesale customer.
- Irrigation water charges, per acre-foot of water supplied (this is given for you).

Discussion. This is a complex problem, but is simplified. The concept of amortizing capital in the problem is that if you borrow to build facilities you must amortize loans or pay interest on bonds and eventually retire them. At the same time, the facilities you build depreciate and lose value. So, you have two cost issues: one to pay back the principal and interest of the borrowed capital and another to set aside funding to renew or replace the facilities when they need further investment. In the simplified problem, you use a single 30-year lifetime to recover capital costs, but you are not considering depreciation and replacement costs very realistically.

20

Water Laws, Conflicts, Litigation, and Regulation

Introduction

It is with good reason that it is often said that water management is "heavily regulated." Every action related to water allocation and water quality management is controlled by some aspect of law. Water conflicts also occur frequently, and their resolution must fall within legal constraints. These realities mean that water managers are often involved with legal questions so that not every issue ends up requiring intervention of attorneys and courts. When attorneys and courts are involved in water issues, water managers should know how to help them find productive solutions. Water manager may serve as expert witnesses or consultants to lawyers in cases. Also, they are often involved with preparing applications for permits and water allocations. These many issues form the background for this chapter.

The chapter outlines fundamental principles of water law and how it governs water management. It tracks a range of topics other than flood related law, which is explained in the next chapter. The topics in this chapter are extensive and include legal concepts, regulatory and administrative law, types of water laws, water allocation law, health and environmental law, instream flow laws, transboundary conflicts, the justice system, and water litigation.

Legal Concepts

Every citizen has some understanding of law because we live in a regulated society. Situations ranging from traffic safety to criminal activity are familiar in everyday life. Water managers require knowledge that goes beyond these basic situations to extend to specific controls on their prerogatives. This knowledge begins with definitions and principles of law.

Water Resources Management: Principles, Methods, and Tools, First Edition. Neil Grigg.
© 2023 John Wiley & Sons, Inc. Published 2023 by John Wiley & Sons, Inc.

Law involves a system of rules and norms to govern society and achieve justice by mediating among different interests of people. It secures compliance by threatening sanctions by the state through enforcement mechanisms. It is implemented through legislation and legal controls of water management procedures.

Water laws have evolved from ancient times along with other laws to balance the rights of people. Ancient examples include how water is shared along water courses, especially in areas where it is scarce. As societies developed, they encountered situations such as allocation of water that required the intervention of some authority to resolve disputes. This was the beginning of institutions like water tribunals, which met to hear disputes about use of water.

The most fundamental legal principle is the Rule of Law, which means that all persons, institutions, and entities must follow society's laws. If the Rule of Law is in operation, you can have a stable functional society, but if it is not, chaos will reign, and people will suffer.

Law can be divided into public and private categories. Public law involves constitutional issues, statutes, administrative law, criminal law, and procedural law. Private law involves civil issues, business law, and labor law, among other topics. Law also varies between federal, state, and local levels, and it can be classified as constitutional, statutory, administrative, and case law.

Water law is at the intersection of different legal categories, such as resources, public health, environmental, and other arenas of law. It has evolved from common law roots through development of rules for water rights, drainage damages, environmental pollution, and other matters.

Regulatory and Administrative Law

Most applications of water law in water management are carried out through regulatory controls. It is important to understand the context of these controls within society and how they work for water management. Regulation is society's way to control undesirable effects of business and other actions, such as highway speed limits, bank regulations, and licensing engineers to practice. The dilemma is that controls are needed to block incentives that create bad outcomes, but people do not like being regulated. So, regulation is needed, but it should be administered with an even hand.

The types of regulations commonly found in water management today stem back to about 1900. At the time, the government had little role in regulating behavior, but regulating interstate commerce became necessary. The arrangements developed through this early experience have extended to regulatory practices across many sectors today. Early water regulations were weak, but they were at least starting out, such as public health regulations.

There are many complaints about the constantly expanding number of regulations. This is a dilemma in the sense that if, whenever a problem is discovered, a new regulation is issued. This creates a ratchet effect of increasing regulations, and sunset provisions are needed to eliminate old regulations when new ones are created.

In the sense of water management, a regulation is a rule or directive made and maintained by some authority, which is normally a statute passed by a legislative body. The United States Code (USC) is the place where laws are published after they are approved as statutes, and the Code of Federal Regulations (CFR) is the place where federal regulations/rules are published.

The laws that cover how regulations are developed and managed is called administrative law. The United States has an Administrative Procedures Act (APA), which is the statute governing how federal agencies develop and manage regulations and how judicial review of them is handled. The Federal Register is a periodic news outlet announcing federal regulatory actions. Systems of administrative law vary among countries and among states in the United States.

Administrative law provides a structure to rulemaking, adjudication, enforcement, and other decision-making. It also addresses constitutional and statutory provisions, court decisions, and directives that define rights and duties before administrative agencies. It also controls administrative procedures and review of actions taken. Features include due process, public and interest group participation, freedom of information, open meetings, and other arrangements for transparency.

The regulatory process implements laws through regulations of government agencies. For each type of regulation there must be an authorizing statute and an implementing agency. The process is generally called rulemaking. There may also be judicial decisions that interpret and/or reinforce statutes.

As an example of rulemaking, USEPA issues a Lead and Copper Rule (LCR) for drinking water. An interim standard was established in 1975, and successive changes were made up through the 1991 rule, which regulating lead in drinking water at the tap. Revisions to the rule have continued as greater understanding about lead in drinking water has unfolded.

Regulatory programs require mechanisms to respond to violations, like law enforcement responses to other violations. Environmental enforcement procedures have developed in recent years, whereas police enforcement has a long history. It is not always clear how to balance incentives and regulatory controls, especially when groups being regulated have a hard time complying. A small water system having problems is an example. When the system lacks adequate financing, is it better to use persuasion or punishment?

Like oversight of police actions, water regulators sometimes need to be regulated themselves. Such oversight extends to all regulation, like medical practice or financial institutions, for example. If regulators are too lax, systems get out of control and harm the public interest. If they are too zealous, they may harm innocent people and systems. Of course, regulators should be at least quasi-independent and not under too much political control. Their work should be transparent, independent, empowered, and effective.

Types of Water-Related Law

In the United States, water laws have been developed to respond to each category of water management. A textbook on water law will normally include these categories, although their names may differ from one author to another. Laws can be classified by branches of government (legislative, judicial, and executive) and include statutes, regulations, executive orders, and court decisions, as well as constitutions of different levels of government.

Laws and regulations of the categories of water management include:

- Water allocation and use (rights in surface water and groundwater, as well as reserved water rights)
- Health and safety (such as safe drinking water and water pollution control).
- Environmental water (such as fish and wildlife protection)
- Rates and charges (such as rates of a private water company)
- Service access and quality (such as adequate water pressure)
- Drainage and flood risk management and legal controls
- Navigation, water-power, and other instream water laws
- Federal agency and other laws governing public organizations

It is useful to classify water related laws by level of government controlling the different types. Most major federal water laws have been established and the emphasis is now on amendments. Important federal water laws are:

- Clean Water Act
- Safe Drinking Water Act
- Federal Power Act
- Endangered Species Act
- Flood Insurance Act
- National Environmental Policy Act
- Authorizations and appropriations for federal water projects

Why some laws are controlled by the federal government, and some are left to states, can be explained by the nation's history. When the Constitution was enacted, state's rights dominated. Now, many changes have occurred, but property rights in water remain state issues. However, the main environmental laws were passed by the federal government. As a reason, one goal of the Clean Water

Act was to equalize conditions among states so that one state could not attract industry from another by offering more lenient environmental laws.

State water laws address issues that are left to them under the Constitution. These include allocation law for surface and groundwater, instream flow laws for water quality and the environment, interbasin transfer law, and operation of public utility commissions to regulate business affairs of private water companies.

Local governments focus on matters of interest to individual communities, like stormwater and floodplain ordinances, control of local groundwater sources, and any type of codes, like water use restrictions, and industrial pretreatment rules.

Law and Water Management

The extent to which law controls water management decisions is illustrated by Figure 20.1. The reservoir has an authorization for its initial construction that controls how it is operated and used. Reservoir releases will be subject to laws such as for instream flows (ISF) and the Federal Power Act (FPA). An interbasin transfer of water may enter the system under control of state laws. Diversions from

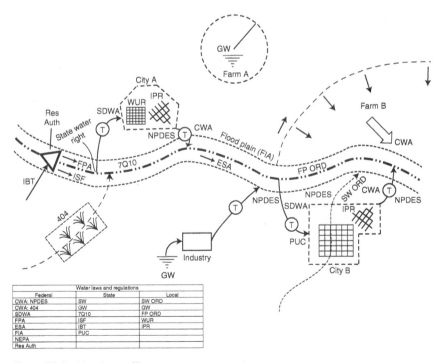

Figure 20.1 How laws affect water management.

the stream are controlled under water allocation law. Also, an industry and farm are pumping from wells and must comply with state or local ground water restrictions.

The Safe Drinking Water Act (SDWA) governs water treatment. Local codes for water use will also influence operation of the distribution system and other water management rules in the city. The wastewater treatment plant (WWTP) must comply with the National Pollutant Discharge Elimination System (NPDES) permit program of the Clean Water Act (CWA). The farm and city must also comply with CWA rules on irrigation and stormwater return flows. The Clean Water Act's section 404 regulations will affect wetlands in a tributary area. Also, 7-day, 10-year low flows (7Q10) rules govern streamflow for water quality purposes. Industrial pretreatment rules govern discharges to the collection system, as well as local ordinances.

In other legal controls, the local flood plain ordinance will govern land use in the flood plain. The Endangered Species Act (ESA) may control water flows and diversions. City B has a private water company that is also regulated by the state's Public Utilities Commission (PUC).

Water Allocation Law

The law of water allocation and use is at the base of many water management decisions. It controls the diversion and distribution of water for cities, industries, farms, and other uses. The water user requires security that the water will be available when needed; otherwise, investments might be made in businesses or farms only to fail with lack of water. This security of water is handled through property systems in water or government actions that provide permits with sufficient duration for the water user.

The need for regulatory control of water allocation through systems of water rights dates back centuries. An example is the Tribunal de las Aguas (or Water Court) of Valencia, Spain, that was in operation before 1000 CE.

Water allocation law is prominent in drier areas. For example, in the semi-arid state of Colorado water might be diverted from a stream for snowmaking in a ski area, a city, a farm, or a factory (Figure 20.2). There may be also a requirement to sustain a certain flow rate in the stream for environmental and other purposes. There may also be restrictions on groundwater pumping to prevent impacts on the stream.

Different systems of water allocation law prevail in different areas, especially considering humid versus arid areas. The main systems are riparian as recognized in humid areas, the appropriation doctrine as practiced in many dry areas, and mixed systems based mostly on permits. In these systems, issues are ownership

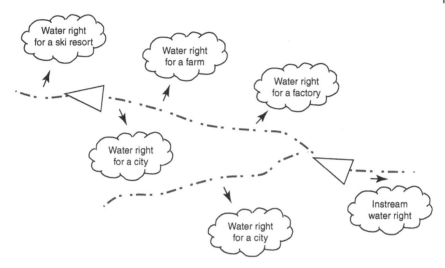

Figure 20.2 Surface water rights allocation in a watershed.

and security of the title, how water rights or entitlements are administered, how measurement and forecasting are accomplished, and how water is to be allocated during drought. Also, certain public trust issues may arise, like how much water should be allocated for the public interest as opposed to private interests.

The Appropriation Doctrine varies somewhat among semi-arid and arid states, but it is generally known as the "first in time, first in right" doctrine. This means that water rights are extended on the basis of seniority or date when the water was first applied to beneficial use. In the United States, its development stems back to the California gold rush of 1848, where the system was developed to apportion water among gold miners. It was extended to Colorado in its 1859 gold rush, and then was expanded to include farm water rights. Afterwards, the system became much more sophisticated and is used to allocate water among all types of rights.

How the system is implemented varies among states, and the Colorado case will be explained as an example. The first step is when a water right is filed, say for someone who wants to irrigate a farm. Colorado manages its system through a system of state district courts, and the water right would be filed in one of these water courts. Not every western US state has water courts, but all have some arrangement to mediate conflicts and handle reallocations of water rights. After an evaluation, the Colorado water court adjudicates the right if everything is in order. The water right owner is required to put the water to beneficial use and show due diligence in perfecting the right. Of course, this is an idealized situation because most water rights that provide reliable supplies were filed and adjudicated more than a century ago. Also, few water right filings would be undisputed.

Administration of the water rights system requires considerable effort in measuring, forecasting, and controlling diversions of water. In Colorado and in most Western states, this work is carried out through offices like the State Engineer. Names of these offices very, but the functions are similar. These officials handle regulation and administration of water rights systems. Their counterparts in more humid states may also regulate allocation of water, often in combination with water quality regulation.

In Colorado, the main activity in the water courts is to examine proposals to exchange and trade water rights. The exchange mechanism for the appropriation doctrine enables the functioning of water markets, which are needed to make water allocation more responsive to societal needs. In these cases, extensive hydrologic studies are carried out to evaluate how the transfer can occur without damaging other water right owners. For example, if a water right from down-stream is transferred to upstream, it may damage other water right owners and require mitigation.

The appropriation doctrine is firm during droughts and allocates water strictly according to seniority. Ways must be found within the system to make things work out to mitigate hardship. Various methods are being developed, such as substitute supply and methods of water trading. The administration of such methods requires drought response plans with triggers and monitoring through metrics like drought indices.

The Riparian Doctrine as used in the United States stems back to common law in England. In that relatively humid region, landowners next to bodies of water had the right to make "reasonable use" as it flows through or over their properties. Reasonable use should ensure that the rights of one riparian owner are weighed fairly and equitably with the rights of other riparian owners. This placed a pre-mium on locating near water sources, and allotments were generally fixed in pro-portion to frontage on the water source. Rights could not be sold or transferred other than with the adjoining land and only in reasonable quantities. Also, water could not be transferred out of the watershed without consideration as to the rights of the downstream riparian landowners.

While the Riparian Doctrine is discussed in legal textbooks and embodied in state laws, it seems to operate more as an underlying concept than as a practical administration tool. Humid states in the United States have tended to trend toward a doctrine which might be called "administrative." Other terms such as "regulated riparian" might also be used, but it is not always clear how important the "riparian" designation is in deciding on rights.

This Administrative Doctrine is primarily based on permits systems. These rec-ognize the reality of resistance to regulation, especially among farmers but also among industries and cities. Generally, permits are issued for water allocations, but they have less impact during wet periods when they are no pressing issues than they do during water shortages. Permits are normally issued for relatively large water users, such as more than 0.1 mgd. Primarily, they become operable during dry periods, especially when water sources become stressed and demands increase during drought.

During these dry periods, water users tend to skim off available supplies and exacerbate water availability for other users and the environment. Therefore, the permits are especially applicable during critical periods. This system helps to promote reservoir construction and implementation of water security measures, as well as to prevent water users from simply exploiting natural supplies when others need them. Of course, drought response plans are needed, and the Administrative Doctrine is more flexible to spread water supplies equitably during drought than the Appropriation Doctrine.

Allocation of groundwater is also handled under systems of water rights, but different methods must be used. Groundwater is the principal source of drinking water for more than 50% of the US population, so how it is managed becomes especially important. While it is a widespread issue, there is no national approach to groundwater management and it is left to state laws.

Groundwater law is a hybrid of water allocation law, resource law, and land use law. The technical aspects are complex, especially due to the different types of aquifers. States take different approaches. In some, little control is exercised over groundwater pumping, while in others sensitive areas are brought under control, sometimes by the formation of groundwater districts.

In groundwater management systems, a state might recognize ownership of the water by the owner of the overlying land but also limit use to reasonable levels. Local ground water ordinances may be implemented by county governments, or management districts.

Stream-aquifer interaction is a concern in administration of groundwater management systems because pumping can deplete stream quantities.

Health and Environmental Law

There are many health and environmental water laws, and a useful way to classify them is as safe drinking water, water pollution control, and other environmental laws. The histories of these laws was described in Chapter 2. Their main provisions are explained here.

The 1974 Safe Drinking Water Act (SDWA) regulates the quality of drinking water based on maximum contaminant levels (MCLs) and health effects. It envisions that utilities will implement a multiple barrier approach to protecting water quality. Utilities sample and report the quality of water that is distributed. Small water systems have unique problems, and they receive special attention from

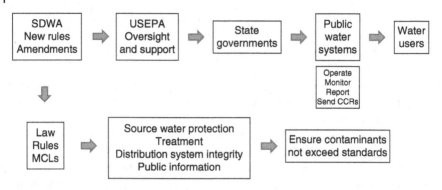

Figure 20.3 How the SDWA works.

special attention from regulators. Plumbing codes control systems located in buildings and other properties.

How the SDWA works is shown on Figure 20.3. The Clean Water Act, passed in 1972, is related to the SDWA and is the most influential regulatory tool for overall water management. How it works was illustrated in Figure 15.4.

The starting point is water quality standards for streams and then monitoring to determine if they meet the standards. If they do not, then they are to be listed as impaired under section 303(d), and pollution allocations are to be made on the basis of point and nonpoint sources. Total maximum daily load (TMDL) studies may be necessary to determine allocation of pollutant loads from different sources. States must develop water quality management plans and implement pollution control reduction strategies. If standards are met, then the antidegradation policy requires that stream water quality not deteriorate.

Point sources are governed by National Pollutant Discharge Elimination System (NPDES) permits. Section 319 of the bill covers nonpoint sources, and stormwater discharges were added to be controlled, along with agricultural and other sources. Section 401 covers water quality certifications, which grant states the authority to deny permits unless conditions are met. Section 404 is the dredge and fill program that provides for assessment of the impacts of any action that may affect waterways. Financial assistance is provided by the Clean Water State Revolving Fund.

Environmental laws that affect water management can be seen as a collection of special-purpose laws that apply to different situations. As explained in Chapter 17, the overarching one is the National Environmental Policy Act (NEPA), which requires environmental impact statements in some situations involving federal actions. These can determine success or failure of plans for water projects, especially when the environmental statements are connected to section 404 of the Clean Water Act. Common situations are those involving dams and actions on streams, as well as any action requiring an NPDES permit.

The Endangered Species Act is also very influential because when a species is designated as threatened or endangered, it can make water development and management changes very difficult. Other environmental laws, such as the Wild and Scenic Rivers Act, can apply in special circumstances. Some other laws have environmental implications as, for example, the Federal Power Act, which requires environmental studies for relicensing of hydroelectric generation facilities.

The Public Trust Doctrine should be mentioned in the context of environmental law. It is a legal principle derived from English Common Law to apply to use of natural resources, such as state waters that are owned by all citizens. It addresses our shared need to manage the environment to serve the broad public interest and not narrow special interests.

Instream Flow Laws

Instream flow laws for surface waters are handled mostly at the level of state governments. Their general purpose is to set aside water in streams for the protection of wildlife and public health. Where they exist, these laws are quite variable in their design and effectiveness. Water quality law that requires minimum instream flows to support flow dilution have similar goals, but there can be the need for additional legal authority under instream laws to coordinate among competing water uses and to avoid stream depletion.

The Colorado instream flow law can be used as an example of a state approach. As the state uses the appropriation doctrine of water rights, environmental instream uses of water were not recognized initially, and this resulted in 100% depletion of waters in some streams. As this was unacceptable environmentally, pressure mounted to provide legal relief such that some minimum streamflows could be maintained. The 1973 law was passed with provisions that it would not injure existing water right owners but provide a mechanism whereby in some situations minimum streamflows could be provided. Lawmakers sensed that environmental interests might seek to file for water rights to have such instream laws applied in a widespread way and injure water right owners, so they applied a restriction that water rights for instream purposes could only be retained by the state's Colorado Water Conservation Board.

Transboundary Water Law

Transboundary issues involving water allocation, water quality, lakes, aquifers, flooding, and environmental water can occur at scales from small watersheds to large international river basins. They involve situations where water moves across the boundaries of political jurisdictions or across watershed boundaries, as in interbasin transfer. There is no special law to govern these, but laws authorizing compacts and contracts can be used.

Transboundary situations can vary in purposes of water management, whether surface or groundwater, and water quality or quantity. The range of possibilities is shown in this list:

- Surface or ground water allocation across borders (such as an interstate compact)
- Water quality issues across borders (such as upstream-downstream pollution)
- A river that forms a jurisdictional boundary (such as between states or nations)
- A lake that straddles borders (such as between states or nations)
- Cross-border aquifers (such as an aquifer that extends from one jurisdiction into another)
- Complex issues crossing multiple borders (such as a river flowing through several nations)
- Interbasin transfers (when basins-of-origin lose water to importing basins)

When water moves across political jurisdictions and exceeds the reach of a single sovereign authority, the legal power to resolve disputes is ambiguous. However, there are types of institutional arrangements that are available to address this problem, and the main challenges are reaching agreement and enforcing compliance.

In the United States, water allocation across a jurisdictional boundary is normally done by interstate compacts and agreements. Perhaps the best-known is the Colorado River Compact, which controls a river that crosses several state boundaries and the US–Mexico boundary. To form such a compact, two or more states must agree and the state legislatures must agree. The authority is in the Compact Clause of the United States Constitution, which provides that "No State shall, without the Consent of Congress enter into any Agreement or Compact with another State, or with a foreign Power."

The United States is fortunate in that disputes between states can be settled by the Supreme Court, but when two sovereign nations have a water dispute, they lack an arbitrating authority, unless they decide mutually to accede to one. In some cases, severe conflicts and even war can result from international water conflicts.

Justice System and Water Litigation

In addition to the executive and legislative branches, the judicial branch of government has an important role in water management. The legislative branch makes laws, the executive branch carries out the laws, and the judicial branch evaluates the laws. Much of water law is statutory, but a great deal is also "case law," where complex situations have been tried and precedents have been set.

Attorneys search for these precedents to prove their points and to build arguments based on them. The water manager should have a basic understanding of judicial processes as they affect implementation of laws and regulations. These processes can involve sources of laws based on the Constitution, statutes, treaties, administrative regulation, and common law.

The starting point of the process that leads to the court decision becoming case law is a lawsuit. When it is filed, it means the voluntary, coordinated approach has broken down, and court decrees and decisions may take the place of agreements and programs. These normally involve complex situations where agreement has been difficult to achieve. The cases provide resources for attorneys, who search for those to prove their points and to build arguments based on precedents.

The justice system involves agencies, establishments, and institutions tasked with administering or enforcing criminal and civil law. Water issues normally involve civil law, but criminal law can be involved as well. The criminal justice system comprises law enforcement, courts, and correctional institutions, and the civil justice system deals with contracts, property, family relations, and civil wrongs causing physical injury or injury to property (torts). For example, in Flint MI episode, injured parties can sue for damages (civil law) and district attorneys/attorney general can ask for indictments to send officials to jail (criminal law).

The justice cycle for a water management issue like pollution, flooding, or water rights deficits begins with people bringing pressure on politicians, who pass laws and appropriate money. The executive branch implements the laws with programs. Disputes arise, which are passed to the judicial system for determinations. The executive branch takes note of the court decisions and changes its implementation. The legislative branch may pass new laws or amend old ones.

The roles in the justice system include plaintiff, defendant, attorneys, experts, judges, juries (in some cases), and mediators. To understand how things work, it is helpful to view the incentives of these different players. The aggrieved party may be aware or not of avenues to receive remuneration. If the party contacts an attorney, that attorney may feel a personal or financial incentive to take on the case. The defendant can be a sovereign, like a local, state, or federal government agency, or it can be a private party. If the attorney hires experts, they have professional and financial incentives to participate. The judges or mediators seek to understand the cases and to make decisions that will stand up under appellate review.

Scientists and engineers may become expert witnesses in cases. Their contributions in federal cases are regulated by Rule 702 of the Federal Rules of Evidence (Testimony by Expert Witnesses). The Federal Rules of Evidence govern evidence at civil and criminal trials in federal trial courts. State courts can follow their own rules.

An expert is a person to give an opinion based on experience, knowledge, and expertise. These opinions should provide independent, impartial, and unbiased evidence to the court. When giving testimony, an expert should state facts or assumptions, and be clear if an issue falls outside the expert's expertise. If an opinion has not been studied well or insufficient data is available, this must be stated. The expert should resist pressure to slant a report toward a party's case and not compromise independence.

Water issues can involve different parts of the court system, and a water issue can be decided at any of these levels, with provisions for review up to the US Supreme Court. These parts include Federal and state trial courts, local courts, administrative courts, and mediation panels. If court decisions are appealed, the cases go to appellate courts. For cases involving federal agencies, like the USACE, the US Department of Justice provides the attorneys to defend the agency actions in court.

There are many examples of legal disputes over water matters. Flood litigation is common, and is discussed in Chapter 21. Other examples include tort litigation about various damages, like pipe breaks or seepage water entering buildings, property disputes over water rights, environmental enforcement actions, and even criminal charges like in the Flint, Michigan, water disaster involving lead in drinking water.

Questions

1 What is the basic purpose of law? How does this apply to law for water management?

2 What is meant by the "Rule of law"? Does it matter in water management?

3 Is water law one of the main topics of the body of law? Why or why not?

4 What types of water laws are dominant for each of the three levels of government?

5 What is administrative law and how does it work in water management?

6 What is meant by rulemaking? Give an example of a federal rule affecting water management.

7 What is the official document where all US laws are compiled? What is the document where regulations are compiled?

8 Would it be more efficient to have regulators be employees of water agencies rather than have them in separate organizations?

9 What is "case law" how does it affect water management?

10 What are the three branches of government and what is the role of each in the water management justice system?

11 Is a Colorado water court a federal court? Explain.

12 What is a tort and what is an example in water management?

13 Which federal court hears cases against the US Army Corps of Engineers?

14 Has the Supreme Court heard any water cases? If so, give an example.

15 When they serve as expert witnesses, do scientists and engineers become advocates for their side?

16 What is a class-action lawsuit and what would be an example in water management?

17 Do courts of appeals try cases all over again or how do they work?

18 Give examples of civil law and criminal law that might arise in water management.

19 Give three examples of how laws control water management.

20 What is the purpose of the State Engineer's Office in Colorado water management?

21 Give an example where water management regulation is needed to prevent some people from harming other people.

22 Are Native American tribes entitled to water rights? How?

23 Explain the concept of Federal Reserved water rights.

24 What is the situation known as "buy and dry" in Colorado water management?

25 Why is water governance more complex or transboundary situations than in single jurisdictions?

26 What are the main US doctrines of water allocation law?

27 Explain how the appropriation doctrine works.

28 How do drought response plans work under the system of Colorado water law? How do they work in riparian systems? In transboundary situations?

29 What is the origin of the riparian doctrine and how does it work?

30 What are permit systems for water use and how do they work?

31 What is the basic difference between groundwater allocation and surface water allocation law?

32 What are interbasin transfers and their legal issues? What is the foundational court case about them?

33 What are the main challenges in achieving effective transboundary water governance?

34 What would be the difference in the political calculation of single jurisdictions and the intergovernmental coordinating group in negotiating transboundary agreements?

35 How is water allocation across jurisdictional boundaries normally handled in the United States?

36 Name some unique features of the management of transboundary Colorado River issues.

37 What is the "Law of the River" for Colorado River management?

38 Would you expect people living in humid areas not to have objections to interbasin transfers? Explain.

39 If Egypt has used the full flow of the Nile River since antiquity, do you consider that upstream countries have rights to take some of the water now? Explain your opinion.

40 What would be the authorizing statute for the Lead and Copper Rule? What would be the implementing agency?

41 In your opinion, should an intentional action leading to a water treatment plant failure carry a civil or criminal penalty?

42 What is the basic US law to protect public health relating to drinking water?

43 What is the basic US law to manage water quality in the environment? What are its main provisions?

44 How does the water quality law control wastewater treatment plant effluents?

45 How does it control nonpoint sources?

46 Is stormwater quality to be regulated under the water quality law? If so, how?

47 What is the concept of the total maximum daily load?

48 How does NEPA work to control environmental aspects of water management?

49 Are instream flow laws at the federal or state level? How widespread and effective are they?

50 What is the legal authority for interstate compacts in the United States?

51 Explain the Doctrine of Equitable Apportionment as it is effective in transboundary situations.

52 Why might a river forming a boundary between political jurisdictions create transboundary problems?

53 What would be the technical difficulty in negotiating water sharing in a cross-border aquifer?

54 In what sense are boundary reservoirs useful in interstate compacts? What do they do to facilitate equity in water-sharing?

55 Give an example and explain the concept of compensatory storage in resolving interbasin transfer conflicts.

Problems

Legal problems in water management are not usually as clear as those involving technical issues. They involve shades of gray in interpretation and may require adversarial arguments and judicial decisions. The six scenarios described below comprise a tour of common water-related issues. Brief discussions are included for each situation.

Water Rights Law

You are advising the water agency of a small developing country in Central America. The government is concerned that during drought periods some people take all the water for their coffee farms and others get none. Unless this situation is resolved, violence may occur. The country operates mostly under the Rule of Law, but administrative agencies are not always reliable. Explain the situation and your analysis of how to set up a legally authorized management program.

Discussion. The country may have a written water law, but implementation is problematic. You may not be able to rely on solution of this problem through regular government channels. Local collective action will be needed, and mediation to get the partners to work together will be necessary. Ultimately, something like a Water User Association may be required.

Transboundary Water Issues

Water flow from Colorado to Nebraska is regulated under the South Platte Compact. http://cwcb.state.co.us/legal/Pages/InterstateCompacts.aspx. The Governor of Nebraska has requested a meeting with the Governor of Colorado because farmers in western Nebraska have threatened a lawsuit over short deliveries during the last few drought years. Your task is to brief the Governor of Colorado on what he might expect to hear at the meeting and what his options are. Explain the major points you would prepare for your briefing to the Governor.

Discussion. The farmers in Colorado, the upstream state, are probably also under stress and it is an issue of sharing the pain. It may be impossible for Colorado to deliver the required water quantities. You should brief the governor of Colorado on how to work with Nebraska and to find a win-win solution that works in the long-term and complies with the compact.

SDWA and CWA

In a hypothetical situation involving permits to operate drinking water and wastewater systems, assume that you are a staff assistant to the director of a combined water and wastewater utility which has just been through a flood and received the following reports: Coliform samples at some water taps have exceedances and you are encouraged to issue a "boil water" notice; the wastewater plant effluent is exceeding permit conditions for biochemical oxygen demand (BOD).

Referring to the SDWA and CWA, prepare an analysis with the following information:

1) Why you are concerned about coliforms in drinking water and excessive BOD in wastewater effluent.
2) How the SDWA and CWA address issues like these.
3) Your suggestions on actions to be taken by the utility.

Discussion. You are concerned about coliforms and BOD because of health and environmental issues. Health issues of the major concern. The SDWA and the CWA address these through regulatory programs and permits specify the actions required of your agency. You would advise your agency to comply with the law, inform your regulators and your customers about the exact situation, and then implement the remedial measures required.

Environmental Law in Water Management

One of the main environmental laws is the National Environmental Policy Act. It has been applied to the ongoing evaluation of a proposed reservoir, which would be built near Fort Collins, Colorado. Your task is to evaluate what the project proponent (Northern Water) must do to comply with the Act. Explain the details as you would brief the board of directors of Northern Water.

Discussion. Complying with NEPA requirements is a challenging part of planning, designing, and implementing a new dam and reservoir. You would brief your Board of Directors on the details of the law, the process required, and the expected stakeholder involvement and opposition. You would outline for them what previous experience showed, and you would suggest the best course of action.

Instream Flow Laws

Elinor Ostrom, a Nobel Laureate in economics, developed theories about how local people can work together through collective action and manage common resources without government regulation. Discuss how her theories (these can be accessed through an Internet search of her work) might apply to two instream flow problems: (i) a relatively small basin, like the Poudre River from its headwaters through Ft. Collins and on to Greeley; and (ii) a large basin, like the Colorado River from Wyoming to California.

Discussion. While Ostrom's theories are powerful, they are very difficult to apply in the complex arena of watershed or river basin management. The difficulty in the larger scales, such as the Colorado River, is that people cannot work face-to-face but must work through representatives who may not succeed. At the smaller scale of the Poudre River, neighbors can work with neighbors and have a better chance of success.

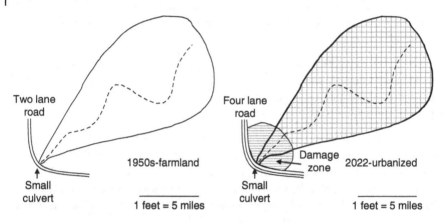

Figure 20.4   Drainage in pre- and post-development situations.

Drainage and Flood Control Law

A common situation in developing areas with land development is changed hydrologic conditions. This can also occur due to climate change. In this case, a watershed has been urbanized since the 1950s. Notice that the small culvert that is still in place is not adequate for larger flows. As a result, a new damage zone appears, as shown in Figure 20.4. In an analysis, outline: the problem, the hydrologic causes, the new risk of damage that has appeared, the liabilities to remedy the situation, and your recommendations for actions on the part of various parties.

   Discussion. Drainage law would say that people causing the problem have a responsibility to pay. The challenge will be to determine how to assess fair charges. For example, if someone develops a site in the upstream part of the basin, they will cause more of an increase in problems than people in the lower part of the basin. Regardless, it will be difficult to differentiate among responsibilities so the best solution is probably some kind of a drainage district with the uniform tax or fee structure across all property owners.

21

Flooding, Stormwater, and Dam Safety: Risks and Laws

Introduction

This chapter discusses the risk management responses that are embodied in the laws and regulatory controls that govern flooding, stormwater, and dam safety. While flooding, stormwater, and dam safety have different legal frameworks, they have similar risk scenarios. Flood risk is of greatest concern due to its widespread and common nature. Dam failures can be catastrophic, but they are relatively rare. Stormwater flooding problems can be major or minor and are focused in urban areas, thus of almost universal concern.

As shown by historic events, floods cause global damages on the order of about $40 billion/year, with annual losses in the United States ranging from $5 billion to $20 billion. The figures do not disclose the heavy trauma faced by people flooded in areas without high property values but with enormous human suffering. Flood damages devastate lives, economies, and public services, and they create trauma and public health issues. Many historic floods are in the history books, like the 1889 Johnstown PA and 1900 Galveston TX disasters. Mississippi River floods include massive events of 1927, 1993, 1997, 2011, and more. In recent years, hurricane-induced floods have topped the news, like Katrina, in 2005, Sandy in 2012, and Harvey in 2017. Hurricane Maria devastated Puerto Rico in 2017 and recovery times were long, with the result that many people went without electricity for months. The worst flood in history occurred in China in 1931 with deaths uncertain but estimates ranging around one million or more.

Flood risk is generally most evident at larger scales like big river flooding and coastal flooding, and flash flooding can also pose high levels of risk. Stormwater can be risky, usually at smaller scales, but its urban effects with major flooding can be catastrophic. Dam safety considerations involve larger floods and the effective functionality of reservoir spillways. While risk management for flooding, stormwater, and dam safety involves similar threats, agency responsibilities, and data sources, the differing legal frameworks and programmatic responses make a unified theory possible only at a general level.

Figure 21.1 provides a conceptual view of how flooding poses varied types of threats. It begins with precipitation, which can produce runoff that enters unregulated streams or streams that are regulated by dams and reservoirs. The precipitation can also fall directly on urban areas and produce stormwater that can be classified as minor drainage or major drainage that borders on flooding. Flood flows from streams or exiting from streams can pass through urban areas and exacerbate flooding there. Dam safety is an issue when reservoirs are used to control flood waters.

Water managers have responsibility to work with others to keep floods from hurting people and damaging property. They must work in the relevant institutional environments to assess the economics of flood loss reduction and how best to use the infrastructures of flood control and stormwater management. They must also evaluate nonstructural programs, including flood insurance, and assess flood program results. To carry out these responsibilities, water managers should understand the scale and seriousness of flooding and the situations that can arise. They should have a good grasp of flood infrastructures, reservoir regulation, and urban detention storage systems. To anticipate threats, they require knowledge of flood risk statistics, data, and forecasting. It can also be useful to be familiar with the evolution of US flood policy and how the non-structural emphasis developed with a focus on floodplain management and flood insurance.

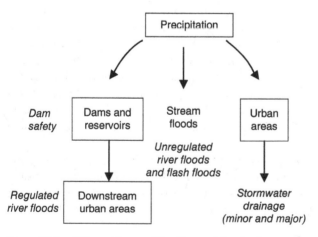

Figure 21.1 Different types of flooding and flood risk.

Flood Threat Causes and Effects

The general aspects of flood risk were discussed in Chapter 7 and they can be summarized in a DPSIR (Figure 21.2). This shows drivers that increase pressures of exposure and risks, which lead to a state of vulnerability and insecurity about the possibility of damages, destruction, and health impacts.

Risk is a function of the likelihood that a threat will occur, the vulnerability of people, property, and the environment, and consequences to them.

The general approach to analyzing flood risk begins with hydrologic studies to determine the likelihood of flood magnitudes. These are then evaluated to assess whether and the extent to which they threaten people, property, and the environment.

Flood risk is such a widespread problem that it has attracted many public and private responses. These include laws, zoning, non-structural solutions like flood warning, and flood insurance. They vary by country, but the general responses to flood risk converges to the same types of approaches.

The concept of floodplain management has evolved to become a framework for a total community-based effort to use non-structural approaches to prevent or reduce the risk of flooding. FEMA has the lead role at the national level to

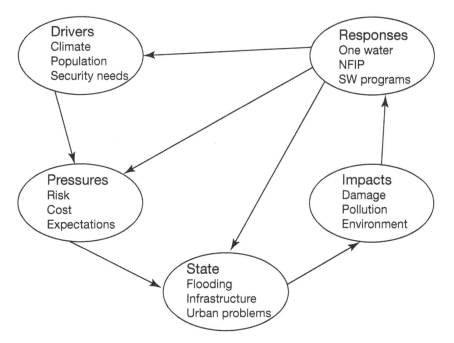

Figure 21.2 A DPSIR of flood risk.

implement tools such as floodplain mapping, flood insurance, local zoning guidelines, education to promote restoration and environmental management, and general methods to reduce flood risk along riparian zones. States have important roles and have banded together and the Association of State Floodplain Managers to cooperate in education and advocacy.

Flood-Related Law

Prior to about 1900, there was no statutory flood law in the United States and any legal proceedings were likely to involve neighbor-to-neighbor disputes. The United States debated the federal role in "internal improvements" such as waterborne transportation, but it did not take action about flood risk. Responses to flooding were focused on individual and local efforts based on levee districts, conservancy districts, and landowner projects.

By 1900, as river and coastal floods causing damage to property and people, the federal government began to recognize that some flood victims might be entitled to relief, but not those who took foolish risks. Flood risk reduction was still considered as a non-federal responsibility for states and their local drainage and levee districts, but the need for a federal role in flood improvements started to gain acceptance.

After 1900, the USACE started projects for river clearing, dredging, and building levees, mainly on the lower Mississippi River and other large rivers. The concept was that levees would induce rivers to scour and maintain channels for navigation and reduce overbank flooding. Despite the levees, damaging floods occurred in the early 1900s, including places without levees or where nonfederal levees were breached. The United States began an era of focus on structural solutions by federal agencies, mainly the USACE. Congress began to authorize flood projects through legislation, like the 1917 Flood Control Act. Federal projects became the favored solution with dams, levees, floodwalls, and flood channels constructed by the Corps of Engineers.

This beginning of federal programs and flood law stemmed from government responses to major disasters, like the 1889 Johnstown, Pennsylvania flood, which killed some 2200 people with the surprising outcome that no one was found liable. The cause was failure of a dam belonging to a fishing and hunting club, with failure due to heavy rainfall. The court found that the flood was an Act of God and assigned no liability. As a result of this injustice, state courts began to adopt a British common-law precedent that a non-negligent defendant could be held liable, which is a doctrine called strict liability.

The 1927 Mississippi River flood was a game-changer. It breached the federal levee system and made a case for a more active federal role in hazard reduction projects with dams and reservoirs. It also spurred the 1928 Flood Control Act, which created immunity for federal government flood control efforts.

Flood protection was added to the progressive-era vision of multi-purpose river development. The Flood Control Act of 1936 established the federal responsibility in flood hazard reduction with the USACE in the lead role for works on large interstate rivers and later in local protection projects. The Act started more projects and initiated the requirement to use benefit–cost analysis. Missouri River floods led to the 1944 Pick–Sloan plan that led to construction of a series of major dams on the mainstem. Control of runoff from smaller watersheds was assigned to the Soil Conservation Service (SCS), which has been renamed the Natural Resources Conservation Service (NRCS).

Despite the momentum in government projects, thought leaders like Gilbert White began to see by the 1940s that large flood control reservoirs were not a good solution. His PhD dissertation was about human adjustment to flooding and pointed to the reality that flooding cannot be defeated and we must adapt to it. This philosophy led to the 1968 Flood Insurance Act, which created the National Flood Insurance Program (NFIP).

The 1968 Act is a public–private partnership where insurance companies can sell policies, which are underwritten by the government. The NFIP includes flood insurance, flood risk mapping, floodplain management, and mitigation. Despite its good intentions, it has generated political controversy and required taxpayer bailouts after major hurricane-induced losses. The flood risk mapping program needs improvement for both riverine and coastal flooding. Reforms have been passed to include basing premiums more on risk, improving mapping, and more. Affordability of premiums remains a major concern.

With reduced emphasis on projects, the USACE budget for flood hazard reduction was reduced. Also, the budget for small watershed projects was reduced and the original SCS program has almost disappeared. The Water Resources Development Act of 1986 shifted more costs of USACE projects to non-federal project sponsors.

After the Great Mississippi River Floods of 1993, a government review committee affirmed the nonstructural approach with emphasis on controlling runoff, managing ecosystems for their benefits, planning land use, and identifying areas at risk. The idea was that many hazards could be avoided and only after that should structural damage minimization approaches like elevation of buildings or construction of flood protection structures be used.

After 2000, it seemed that the flood hazard worsened, especially with damaging events like Hurricane Katrina that devastated New Orleans in 2005, along with other intense hurricane events that caused widespread damage and destruction. The flood insurance fund required a bailout and, although some reforms of the NFIP were passed, many improvements in it were still needed. For example, many flood damages have occurred outside of designated flood plains, which calls into question the validity of setting insurance rates based on perceived risk.

As a legacy of its construction program, the USACE has about 700 dams to manage risk of flooding and many miles of levee systems, which are managed and coordinated through its Office of Civil Works. The dams are operated according to USACE Engineer Regulations. The levees present a different challenge as they must be maintained in coordination with a number of local levee sponsors and stakeholders. In addition to dams and levees, the USACE manages an extensive inventory of river training structures.

While flood policy and statutory law address some aspects of flood risk, damaging events may cause people and organizations to seek relief through the judicial system. This is anticipated through the general tenets of common law, and it has been handled through the court system for decades. The common law as explained later in the stormwater discussion is more amenable to small scale events, whereas larger cases will be considered on the basis of various facets of liability. A starting point for the discussion of liability could be the Johnstown, Pennsylvania, flood of 1889 where, despite death and damage, no one was held liable and the injustice led to development of the doctrine of strict liability in the US legal system.

The USACE flood infrastructure has reduced flood hazards in many areas, but it can increase damages if floods surpass the capacities of reservoirs or breach levees. The infrastructure is aging and there are many concerns for its reliability, structural integrity and degree of protection provided. As it manages this infrastructure, the USACE faces challenges to provide project benefits that may be inconsistent due to authorizing legislation and related regulations, as well as natural limits of ecological systems. USACE actions for reservoir operations, river training, and levee management can all be challenged.

Many flood cases involve individuals and local organizations, and a number of high-profile flood problems associated with federal projects have also occurred. To outline the range of scenarios that may occur, the following list provides examples from recent flood cases, mostly from those relating to federal projects and actions:

- People build in vulnerable areas and larger-than-anticipated flood occurs
- Levees fail or are overtopped
- Bridge restricts flows and causes upstream flooding
- Flood gates on a reservoir opened rapidly and increase flows downstream
- Operations rules change and flooding worsens
- River training structures modified and worsen flood conditions
- Dam may fail and damage downstream property

Flood lawsuits for situations like these can be tried in courts from local up through the federal court system including the Supreme Court. While the federal government is immune to liability claims for flood control projects, it can be liable for negligence in projects without flood authorization purposes. The 1928 Flood Control Act established immunity from tort liability for flood damages but not takings due to government actions.

Takings litigation refers to cases based on the Fifth Amendment to the US Constitution that restricts government taking property without just compensation. Sometimes this is called "inverse condemnation," which means a situation where the government takes private property or part of its value but fails to pay compensation, and the property owner must sue to be receive just compensation. Some interesting legal precedents have been established in flood cases related to it. For example, in a 2012 flood case, the Supreme Court ruled that government is not obligated to help with social goods, but the Constitution restrains it from depriving basic rights like not having property damaged due to flood projects. Takings cases do not require that negligence is proved, but Tort liability cases do require it.

Flood cases alleging either negligence or takings illustrate several important legal issues. One is intent and was an action deliberate and intentional? Another is cause, meaning was the action a direct, natural, or probable cause of the action, like the raising of a spillway gate? The issue of foreseeability is a test of whether the damage could be foreseen at the time of the action. In terms of degree of damage, questions include how intrusive was the action and whether it caused temporary or permanent damage. Another issue is whether it is likely that the damage will recur. Reasonableness is another test of whether the action was unlawful and/or unreasonable. Finally, there is a question as to whether the government benefitted at the expense of the owner.

Stormwater

Stormwater issues are different from major river and coastal flood issues. As stormwater management systems evolved, they passed through phases of only draining streets to a new scope of draining growing subdivisions, and then on to environmental and water quality goals. Because they have multiple purposes, the risk factor for stormwater is not as prominent as for larger-scale river flooding, but stormwater flooding in urban areas can be serious as well, especially when major drainage pathways in cities are blocked. Minor drainage and major drainage in cities carry different levels of risks, and failure of stormwater quality control programs creates environmental and health risks. Minor drainage does not normally lead to as much property damage as major flooding, but problems like flooded basements, backed up sewers, and transportation problems can be significant.

Responses to stormwater risks generally take the form of comprehensive programs by cities. These provide programmatic responses for minor and major drainage, floodplain management, stormwater quality, and enhancement of open space and recreation. Failures in these programs can lead to unacceptable damages and costs, and if the programs are not managed well, the result can be lost opportunity to enhance cities.

The main risk category for comprehensive stormwater programs is in drainage and flooding, but water quality and urban development should also be considered. Stormwater quality became a program objective during the 1970s under the authority of the Clean Water Act. Stormwater is a nonpoint source of pollution, and the program today is called the MS4 for Municipal Separate Stormwater Systems program.

The risks are mainly in the regulatory category, although stormwater contamination can pose serious problems for receiving waters and the environment. The emphasis of stormwater on urban development focuses on green infrastructure, LID, BMPs, rooftop gardens, and similar measures.

Stormwater law mirrors the different types of services for minor drainage, major drainage (urban floods), stormwater quality, and urban development. The law that governs minor drainage involves property owner rights and responsibilities. This is sometimes handled under the law of "diffused waters" and is codified in local land use law that varies by states. Relief is provided to injured parties through on a case-by-case basis rather than by direct regulation.

The main drainage law doctrines are the common enemy and the civil rule doctrines. In the common enemy doctrine, each landowner is responsible to protect his or her land, while in the civil rule a landowner must protect others by not discharging damaging levels of drainage water. A modified civil rule or reasonable use doctrine is sometimes used as a mixture, and with it there is liability only if land use is unreasonable. Many issues arise that require legal decisions, such as when a town improves a street causing homeowner damage, when a shopping center increases paved areas, when groundwater due to lakes and ditches floods homes, or when a city wants to condemn a property to open a route for a drainage system.

An important challenge with stormwater is how to pay for management programs. State laws provide the legal structure financing models through tax-supported public works or fee-based utility models. As a result of tax limitation initiatives, a stormwater utility model was created to garner revenues for stormwater programs using the fee approach. This approach has been challenged in courts, but the stormwater fees have been upheld in a majority of states. They must be reasonably related to value of services, funds not used for general revenue, fees structured to contribution of runoff, and people must be able to choose to provide own systems or offset volumes of runoff with various credit schemes.

Dam Safety

Dam safety leads to flood risk, and it is a critical concern due to the extreme threats to people and property. As explained in Chapter 2, the United States alone has some 90 000 dams that are considered large enough to monitor, and many of them pose significant risks. Dams can fail due to overtopping, structural failure, cracking caused by movement and settling, inadequate maintenance, and piping, or leaks through the dam. Historically, many dams have failed, such as the Johnstown, Pennsylvania, failure in 1889 that was described earlier.

The US dam safety program is a partnership between the federal and state governments and other stakeholders. Federal dam owners, especially the USACE and USBR, have dam safety programs, and almost all US states have dam safety programs that are authorized by their state laws. This was not always the case and many state programs were spurred by federal legislation.

The National Dam Inspection Act of 1972 established the National Inventory of Dams maintained by USACE. It came after the 1972 failure of Buffalo Creek Dam in West Virginia that caused 125 deaths. Other significant failures of the 1970s included Teton Dam, a USBR dam in Idaho, that failed in 1976 during its first filling as a result of a leak that opened near the northwest abutment. Another significant failure was the 1977 failure of Kelly Barnes dam in Toccoa Falls, Georgia, with 39 deaths.

FEMA has been assigned to coordinate federal dam safety efforts and has a good bit of influence on state programs. It coordinates a program to classify dams according to safety. Many "unsafe" dams in the United States have been identified, along with more state-regulated "high-hazard" dams, which are dams that threaten loss of life if they fail. An unsafe dam will have deficiencies that leave it susceptible to failure and high-hazard dams reflect increasing downstream developments, which place more people at risk.

The Association of State Dam Safety Officials (ASDSO) was organized to provide for cooperative programs among the states, and it now has some 3000 members. According to Charles Gardner, who was president of ASDSO in 1986–1987, the Association was formed in 1984 to provide a forum for state agencies. In the 1970s, federal agencies had improved their dam safety programs in response to dam failures and passage of the National Dam Inspection Program (PL92-367) in 1972. This left many non-federal dams as responsibility of the states. A US Committee on Large Dams (USCOLD) publication in 1970 of a Model Law for State Supervision of Safety of Dams and Reservoirs stimulated interest and a 1982 report of the National Research Council Committee on Safety of Non-Federal Dams showed the need for better state dam safety programs. As of 1982, over half the states either had no dam safety law or no dam safety program. Citizens are often unaware that living in flood-prone areas due to the risk of dam break, and dam owners may not know their full responsibility and liability. Therefore, state efforts to identify and publicize high hazard areas are critical.

Questions

1   What is the approximate range of average annual dollar losses in the United States due to floods? Do the statistics capture the full extent of losses?

2   What were the general causes and results of the 2005 Katrina flood disaster in the United States?

3   What is the main difference between a big river flood and a flash flood?

4   Explain the allocation of responsibility for flood risk management among governments.

5   What is the significance of Gilbert White's work to flood policy?

6   What was the long-term legal impact of the 1889 Johnstown flood?

7   What are the main features of the 1917, 1928, 1936, and 1944 Flood Control Acts?

8   Who bears responsibility to pay for flood proofing of properties?

9   What are the main problems with the NFIP?

10   What is a "takings" case and what is a "tort case" relating to flood losses?

11   What are the main doctrines of surface drainage law?

12   When did national attention to dam safety become intense and why did it occur?

13   About how many dams are in the US inventory?

14   Relative to flooding, what are emergency plans and emergency operations centers?

22

Water Security: Natural and Human-Caused Hazards

Introduction

Water managers are confronted with an array of threats that stem from natural and human-caused hazards. The results can include failures, disasters, emergencies, accidents, and malevolent attacks, and measures to address them can be lumped into the general category of water security. The term water security has taken on broad meaning as a concept to encompass everything that might threaten our access to water and uses of it for various purposes, as well as our protection against harm.

This chapter addresses risks that stem from all threats to water systems, along with principles of planning and preparedness that can strengthen the systems and facilities. It explains the threats, why they are important, and what to do about them. Natural and human-caused threats are discussed, except flooding, which is discussed separately in Chapter 21. Water system managers can use this information to think beyond system design, construction, and operation and add security and risk management to their organizational responsibilities to make preparedness and security core business processes.

Water Security as a Concept

As threats to water have increased, it has become useful to use the term "water security" to organize them under one banner to focus on total risk. One accepted definition of water security (by UN-Water) is "The capacity of a population to safeguard sustainable access to adequate quantities of acceptable quality water for sustaining livelihoods, human well-being, and socio-economic development, for ensuring protection against water-borne pollution and water-related disasters, and for preserving ecosystems in a climate of peace and political stability." Definitions like this become broad because they expand to include a range of topics. By breaking this one down, we see that water security has six elements:

Water Resources Management: Principles, Methods, and Tools, First Edition. Neil Grigg.
© 2023 John Wiley & Sons, Inc. Published 2023 by John Wiley & Sons, Inc.

- Adequate quantities of water – Fair water allocation and effective drought response
- Acceptable quality of water – Safe water for drinking and other uses
- Water pollution – Management of environmental water quality
- Preserving ecosystems – Sustain ecosystems and their services
- Disaster protection – Address floods, dam safety, pipeline failures, and other disasters
- Climate of stability – Achieve effective water governance to address threats

The six elements include water quantity and quality, both in the environment and in water supply systems, environmental management, response to disasters, and water governance. In addition, water management organizations will face other business risks. These can include financial and business risks such as financial shortfalls, enforcement actions, and employee issues relating to health and safety or job security. Cyberattacks are becoming more likely. Sometimes the threats stem from actions by others, such as industrial waste contaminating an aquifer or an industry dumping toxic waste in a drainage system. By the nature of its business, a utility can face construction risks such as system damage by other parties, trenches caving in, and pipes damaged during construction, among others.

These threats have increased as society has become more complex (Figure 22.1). Prior to the mid-nineteenth century, most risks to water organizations were financial and political as water supply services were simpler and cities had not yet initiated wastewater or stormwater services. Public health risks were large and waterborne

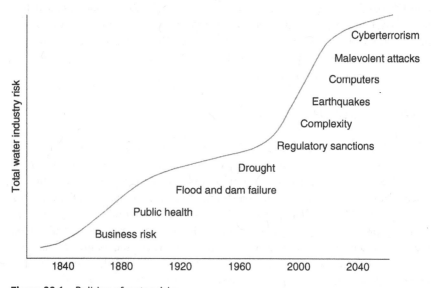

Figure 22.1 Buildup of water risks.

diseases were common, but the public health system had not developed in response. People were mainly on their own to manage risk. Dam failures occurred, such as in the 1889 Johnstown, Pennsylvania, flood that caused many deaths. As cities developed and farming expanded, drought risk became higher. More complex water systems were subject to earthquake damages, and the Chicago fire of 1871 and San Francisco earthquake and fire of 1906 had exposed their vulnerabilities. With the environmental movement, regulatory sanctions and new political conflicts increased risks to water managers. Computers made systems more vulnerable to failure and hacking, and cyberattacks have become more common. The possibility of accidents and deliberate malevolent attacks seems to be on the rise.

Threats and Impacts by Category

Security and disaster preparedness officials are confronted by multiple threats, which can be classified in different ways, such as natural, human-caused, and hybrid. Natural disasters can be biological, geophysical, meteorological, hydrological, climatological, and extra-terrestrial (like meteors). Human-caused disasters can include accidents, malevolent activities, civic disorder, and others. Also, threats can be direct and indirect, such as from interdependence with other systems. There are so many types of threats that a full classification system becomes burdensome, and it is more productive to focus on those that pose the greatest threats to water security.

Using natural hazards and human-caused (non-intentional and intentional) as basic divisions, the list below shows disaster categories that seem to pose the greatest threats. This makes a long list and illustrates the complexity of preparing for all eventualities.

Natural hazards	Atmospheric	Lightning
		Windstorms
		Ice and snowstorms
	Geological	Earthquakes
		Tsunami
		Landslides
		Subsidence
	Hydrologic	Floods
		Droughts
		Wildfires
	Biological	Epidemics

(Continued)

Human-caused	Non-intentional	Accidents
		Technological failures
		Hazardous material release
		Fires
	Intentional	Mass shootings
		Civil disobedience
		Terrorism
		War

Each type of threat presents different risk factors that depend on vulnerability and potential consequences to water systems and their stakeholders. Natural hazards such as floods, earthquakes, and wind are addressed by the HAZUS Multi-Hazard program of FEMA. System failures can be due to infrastructure and equipment and from interdependence with electric power and computer systems. Waterborne disease outbreaks are difficult to anticipate, as are attacks by actors with malevolent intent that include terrorism, vandalism, sabotage, hoaxes, and cyberattacks.

Civil conflicts can include riots, strikes, and general unrest, as well as wars that create refugee crises and water systems themselves.

Drought and Climate Shifts

Droughts and climate shifts can create water insecurity and lead to refugee crises that overload water systems. Drought threats vary by time of onset and ending, duration, intensity, and extent. In contrast to flooding, drought is a creeping disaster rather than a sudden disaster, and it is difficult to forecast its onset or duration. Drought scope, timing and intensity are difficult to describe. Drought varies geographically, and maps and statistics are required to characterize it.

Meteorological drought causes lags in hydrologic systems and shortages in urban supplies that can extend past the periods with lack of rain. Water managers use drought indices to measure drought intensity as the difference between needed and available water resources and possibly to trigger water use restrictions. Safe yield, which was explained in Chapter 6, is a useful tool to measure the capability of systems to meet demands despite drought. The Palmer Drought Severity Index (PDSI) is a measure of meteorological drought that considers precipitation, evapotranspiration, and soil moisture. A national map of PDSI values is published monthly in the USDA Weekly Weather and Crop Bulletin.

The primary impact of drought is from real or feared interruption of supplies. Polls have shown that people's greatest concerns about water are about shortages and contamination. Drought has also been shown to have serious effects on ecosystems and leads to damaging wildfires. Drought threatens food supplies and farm income, and it creates hardship in rural areas. Drought-induced fires show the interdependence among systems. Drought causes dryness and lack of moisture. Fire starts in the dry systems and damages forests and ecosystems. Rain then causes excessive erosion, filling streams and reservoirs with sediment, further damaging ecosystems. Eventually, the forests recover, but if too much soil has been eroded, long-term damage may occur.

The impact of drought on sectors of the economy varies and is hard to quantify and drought management can be difficult politically. It requires intergovernmental coordination and can involve conflict between interest groups and regions. Drought responses are complex when there is no single management agency and unity of command is lacking. The roles of state governments are important to provide coordination among agencies and interest groups. State governments can also provide data and technical assistance, emergency aid to local governments and agriculture, and regulatory actions, mainly restricting water use. The Federal Government has important roles in operation of federal reservoirs, in coordination and in data management.

Better planning, data, study techniques, along with coordination, collaboration and communication will improve drought water management. Water supply agencies should be self-reliant but balance it with collective security of water supplies. They can improve security by gaining access to independent sources. The variability of the sum of independent sources is less than the variability of each one separately. More water storage and management capability may in some cases be needed.

Risk Management and Disaster Preparedness

As new threats emerge, water managers must think creatively. Because water, wastewater, and stormwater utilities cannot tolerate much risk, they must assess the likelihoods and consequences of threats to create plans to manage resulting risks. Preparing water systems to be robust and strong, resilient and able to recover from shocks, and reliable so they do not stop operating requires definite processes, which can be called security programs, disaster preparedness, or emergency management, among other names such as business continuity planning or contingency management, for example. These broad and overlapping concepts share a core set of processes that revolve around risk management.

Risk management is practiced across public and private fields ranging from utilities and public systems to businesses and social systems. As a general concept, risk involves likelihoods that threats can succeed, vulnerability of the person, facilities, or systems involved, and the consequences that occur when a threat is successful. Risk management organizes and coordinates activities to direct and control an organization with regard to risk. It involves a range of programs and concepts that describe preparation and response and that include safety, security, and emergency management programs. Activities range across different phases for preparedness, mitigation, onset, response, and recovery. In the response phase, the Incident Command System (ICS) is widely used, and the Department of Homeland Security has created a National Incident Management System (NIMS).

Tools to conceptualize risk include risk triangles and risk matrices. In a risk triangle, likelihood, vulnerability, and consequences are at the three corners. If any corner moves due to changes in likelihood of threat, vulnerability, or consequences, then risk changes. The risk matrix shows how to classify risks by likelihood and consequences. The variable of vulnerability is implicitly included in likelihood, where vulnerability increases exposure of a target, and in consequences, where vulnerability opens the target to greater impacts.

To illustrate a risk matrix, the display below shows four quadrants of hypothetical threats to an urban water supply system. The entries are dependent on the local situation but they illustrate the concept. No threats are shown in the upper right quadrant because, as noted, such catastrophic events will be recognized and mitigated until their probabilities are reduced. In the upper left quadrant, threats that must be mitigated but are impossible to control totally represent serious concerns. For example, an earthquake or cyberattack can cut off water service for a long duration. A dam failure to eliminate water storage as well as threatening downstream residents. In the lower left category, a plane crash could have serious local effects but be unlikely to cause systemic failures. A failure in the SCADA system can normally be mitigated by instituting manual controls. In the lower right quadrant, frequent events are normally handled as they arise and do not represent serious systemic issues.

High impact – low probability incidents Earthquake Dam failure Cyberattacks	High impact – high probability incidents
Low impact – low probability incidents Plane crash SCADA failures	Low impact – high probability incidents Power outages Broken water pipes Vandalism

Within a water organization, everyone has some responsibility to control risk, but a single department and manager may have responsibility to coordinate overall risk management, safety, security, or emergency preparedness. Figure 22.2 shows a hypothetical the utility risk management organization with one line of activities focused on facilities and operations and the other on organizational

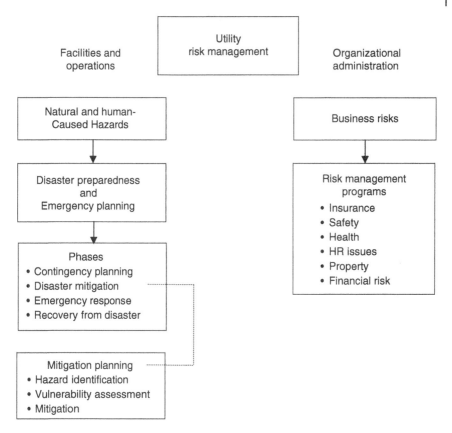

Figure 22.2 Risk management in a water organization.

administration. This would normally be a staff function relying on help from different parts of the organization to develop a combined and coordinated approach.

Utilities need a top-down and bottom-up approach that fits within their overall risk management program. While they may have safety, security, and emergency response programs, they must integrate them through organization-wide efforts led by top utility managers, who can fight complacency. Organizational commitment and leadership require a champion at the top to fight complacency and guarantee commitment. Managers must realize the need and create a plan, employees must cooperate and partners must be involved. A utility may have an emergency plan, but the plan can be useless unless the organization is prepared well.

Organization-wide efforts should include: a focus on planning and coordinating; a living emergency plan to provide cross-cutting coordination during emergencies; plans to adapt to rapidly unfolding unstructured decision scenarios; training for emergency planning; plans for intra- and inter-organizational coordination; use of drills and exercises to transfer lessons learned; matrix management to create lines of authority for emergencies; use of an "Incident Command System"; and study of varied scenarios, including those involving interdependencies between systems and components.

Because disaster preparedness and emergency management are organization-wide activities, best practices go beyond responsibilities of individual utility departments. Some of these are: maintain an emergency plan with an all-hazard approach; prepare for complexity as disasters can only be overcome with serious preparation; maintain a clear line of command with planning and preparation, including drills; mobilize the organization with ongoing attention with plans and exercises; mitigate with facility redundancy and back-up power systems; use pre-positioned repair systems; communicate with good relationships internally and with external organizations that might provide mutual aid; coordinate and cooperate with committees, staff meetings, clarification of roles, cross-communication, and project-oriented units; conduct drills and table top exercises; and assess preparedness. Emergency preparedness must also be funded and records should be maintained during emergencies as well, especially during flood and fire.

Disaster preparedness and emergency management depend more on people than anything else. If water utilities are well staffed and led, and if key personnel know their jobs, they can prepare for and respond effectively to contingencies. Institutional memory is very important as history shows us that disasters will continue to occur in the future. Adaptable and well-led organizations that learn and use good management practices will survive and continue performing their core mission of providing reliable and safe water.

Terminology of Risk Management

The terminology used in risk management is often defined independently by various groups, but an ISO standard (31 000) has been developed to clarify terms and processes.

Key terms that may be useful to water managers planning for risk to their systems are given here in brief form. These may differ from those in ISO 31 000 and other guides, which are modified from time to time, but they provide the main concepts and illustrate the language of risk management.

Term	Definition
Contingency planning	Planning to assure continuity between the impact of a problem and return to normal functioning of an organization.
Consequence	Outcome of an event.
Crisis management	Special measures to solve problems during a crucial or decisive point or situation.

Term	Definition
Disaster	An event that will cause widespread destruction and distress. The term is used in medical, search and rescue, relief, fire, and accident response.
Emergency management	The term is used for police, fire, and medical work, as well as by utilities. The Federal Emergency Management Agency (FEMA) uses the phrase in its name. "Emergency preparedness" is also used.
Event	Occurrence of a particular set of circumstances.
Hazard	A potential source of harm or cause of potential damage from a disaster. See also "threat." A hazard does not lead to disaster unless it penetrates the defense.
Incident management	Management of an event of a certain type. A widely used concept is the Incident Command System (ICS).
Likelihood	The chance or probability that an event causing consequences will occur.
Mitigation	Limitation of the likelihood of negative consequences.
Recovery	A phase of an emergency after response until systems return to normal or better.
Response	The action following an emergency or disaster.
Risk	Combination of the probability of an event and its consequence.
Risk analysis	Systematic use of information to identify sources and to estimate the risk. Risk assessment is a related term with a slightly different meaning.
Risk management	Coordinated activities to direct and control an organization with regard to risk.
Risk treatment	Process of selection and implementation of measures to modify risk.
Safety and security	Safety and security mean about the same thing, freedom from unacceptable risk due to harm, danger, and risk of loss.
Source	Item or activity having a potential for a consequence.
Terrorism	The unlawful use of force or violence against persons or property to intimidate or coerce a government, the civilian population, or any segment thereof, in furtherance of political or social objectives (FBI definition in Code of Federal Regulations).
Threat	A potential source of damage. Similar to hazard.
Vulnerability	The level of system susceptibility to consequences from possible threats.
Vulnerability assessment	Estimations of the levels of vulnerability of systems and the users of the systems.
Vulnerability analysis	Process to identify possible vulnerabilities.

Vulnerability Assessment

As greater recognition of emerging risks has emerged, the process of vulnerability assessment has become a useful and frequently used tool in risk management. Vulnerability analysis generally refers to an examination of the elements of something, whereas assessment generally means the estimation of the level of something, like an assessment of a stream's water quality. Thus, risk assessment is the favored term of the two. Analysis is a precursor to assessment and helps us understand, while assessment tells us what things mean.

In vulnerability assessment, the effects of the threats on water system components are determined. The basic steps and tools to help accomplish them are shown in this table:

Steps in vulnerability analysis	Tools available
Identification of system components	Maps, GIS Databases of system information System diagrams
Quantifying disaster magnitude and effects	Design basis threat concept Threat probability Scenario development Expert opinion Damage-probability matrices
Estimating demands and system capability during emergencies	Demand forecasts Simulation models Expert opinion Charts of data and performance
Identifying critical components	Prioritization based on criteria and weighting factors

Mitigating Risk

After vulnerability assessment, the next step is planning for mitigation, protection, and response. Mitigation consists of disaster-proofing activities which eliminate or reduce the probability of disaster effects to limit vulnerability of components or systems. Planning for mitigation requires identification of protection and preparedness program options and evaluation of options using economic, social, political, legal, and financial criteria.

Mitigation strategies include management and engineering measures. Examples of management measures include emergency response planning, organization of administrative systems during emergencies, cooperative plans for mutual aid, improvements in communication and controls, back-up systems, personnel cross-training, assuring safe workplaces, and placement of redundant equipment and auxiliary generators. Engineering methods include strategies such as providing alternative sources, strengthening and retrofitting water infrastructure, removing high-risk components, seismic upgrades like flexible pipelines and joints, moving systems away from high hazard areas like landslide zones, and repairing leaks in areas of unstable soils.

The basic mitigation of threats of attack is the security system. Government and industry associations like AWWA have prepared documents and trained utilities on security and many principles and lessons learned are available. AWWA has published an emergency planning manual, which is updated periodically. The experience base is also growing and shows minor incidents of attacks or threats of attacks, and prospects of major chemical or biological attacks against potable water systems. Vandalism and sabotage have been reported, ranging from teenagers writing on water tanks to break-ins at treatment plants. Cyberattacks have been detected, and utilities are concerned about vulnerability of SCADA systems. So far, no major act of bio-terrorism or chemical contamination has been detected but the threat of direct attacks remains.

The US Department of Homeland Security has water sector plans, including dams, along with water, wastewater, and stormwater infrastructure. As a connecting resource, water is a factor in plans for other sectors as well, including public health and energy, among others. Threats to water utilities are reported through an Information Sharing and Analysis Center (ISAC) for the water supply sector. The WaterISAC is a secure web-based interface for water utilities to share security information.

Questions

1 Give examples of natural hazards and human-caused hazards.

2 Explain the concept of "water security" as defined by the UN.

3 Give examples of events causing failures: system performance, infrastructure failure, hazardous material spills, and malevolent events.

4 Explain how cyberattacks occur and pose threats to water systems.

5 Identify and explain concepts for the field of risk management such as emergency management, contingency planning, etc.

6 Explain how the United States has organized its response to water system security.

7 Define robustness, resilience, and reliability in the context of water system.

8 Draw a risk triangle and explain its parts.

9 Give an example of how threat, vulnerability, and consequences can vary to increase total risk.

10 Explain the concept of interdependency and how it affects security of water systems.

23

Integrated Water Resources Management

Introduction

With its many diverse situations involving multiple objectives and stakeholders, water resources management requires collective action combined with comprehensive and coordinated strategies. It is difficult to package these broad requirements into a single phrase, but the one that seems to have most current acceptance is Integrated Water Resources Management (IWRM). This chapter explains IWRM as an integrative framework and offers examples to illustrate how it works.

IWRM was introduced briefly in Chapter 1. While there has been much discussion about it, it seems to mean "good water resources management," which is a phrase used by the Past President of the World Water Council, Ben Braga. Said another way, if water resources management is effective, what is left to integrate? Of course, the broad language inherent in words like "management" and "integrate" opens the discussion to different interpretations.

A model that makes sense to me has water resources management at three levels. At the lowest level, it involves basic operations like operating pumps and turning valves. At an intermediate level, it extends to considering various technical and management issues to make decisions about water allocation, diversion, treatment, and other processes. At the highest level, where IWRM resides, it involves considering various stakeholder views, political issues, and regulatory and legal constraints, while also evaluating many ways to meet needs or solve problems. It seems obvious that any water issue involving uncertainty or conflict will require such an IWRM approach.

This three-level framework for water resources management would seem to be practical and to align with the most-accepted definition of IWRM by the Global Water Partnership: "… a process which promotes the coordinated development and management of water, land and related resources in order to maximize

Water Resources Management: Principles, Methods, and Tools, First Edition. Neil Grigg.
© 2023 John Wiley & Sons, Inc. Published 2023 by John Wiley & Sons, Inc.

economic and social welfare in an equitable manner without compromising the sustainability of vital ecosystems and the environment." This definition is general and does provide a good view of how IWRM works. In fact, while IWRM may be defined as a process, it is not a process that can be initiated and then put on autopilot. It requires sustained leadership and coordination, as well as to use effective management practices like data management, modeling, and program assessment.

How IWRM Works

While it is general and seems vague, the definition by the Global Water Partnership (GWP) is based on much input and thought, and it can be parsed to explain how IWRM works. This is implicit in the statement that IWRM should promote the "coordinated development and management of water, land and related resources." Boundaries on IWRM actions are provided by the constraint not to compromise "the sustainability of vital ecosystems and the environment." These broad statements leave much to the imagination as they frame water situations that range across the purposes and scenarios explained elsewhere in the book. Although maximizing welfare, using coordination, and recognizing environmental constraints are not new practices to water resources managers, the IWRM framework helps to assemble them into a set of useful best practices. These best practices are discussed in detail in the IWRM "Toolbox" assembled by the Global Water Partnership and can be accessed by searching under that name.

The IWRM framework can be used to create a model of the management process that applies across different situations (Figure 23.1).

The model is shown as being linear for simplification, but it is not linear because the steps involve feedbacks and adaptation to changing perceptions among stakeholders. They begin with detection of some situation that calls for joint action. The IWRM approach then emphasizes early stakeholder involvement to develop a collective understanding of the problem. Technical studies are conducted to develop alternative solutions, and these must be analyzed for management issues such as risk, finance, and sustainability. The solution requires decisions based on some degree of consensus among stakeholders. This framework is consistent with

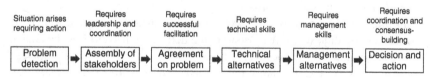

Figure 23.1 Water resources management process with descriptions.

concepts for general problem-solving, decision theory, water resources planning, and the systems approach to defining and solving problems.

Situations Requiring IWRM

The concepts of a problem space for water resources management and how water problems follow common patterns were explained in Chapter 1. IWRM works within this problem space for the problem scenarios involving multiple issues and stakeholders. Problems involving fewer issues and people can normally respond to more structured solution approaches.

Water issues are found everywhere and in many different forms. The common scenarios presented in Chapter 1 provide a way to classify them, but other ways can be devised as well. In any case, the scenarios offer us a way to explain why IWRM is needed in different types of situations. For example, policy development requires us to work at different governance scales and will draw in many stakeholders and participants in decision-making. Watershed and river basin management require alignment of the goals and plans of stakeholders with common and conflicting interests. Infrastructure development goes beyond the engineering space to require analysis of demands, alternatives, feasibilities, trade-offs, and impacts. Other examples can be provided for operations, regulation, and conflict management, which are common IWRM scenarios.

Case Studies of IWRM in Action

This section presents several case studies to illustrate the requirements of the IWRM model shown above. The goal is to distinguish how IWRM should work in water management situations by showing the problem attributes that include appropriate situations, required leadership, facilitation, technical skills, and management skills, along with coordination and consensus building during the process. Many other cases could be presented, but these should suffice to distill the main lessons about IWRM. The writer has some involvement in or detailed knowledge about each case.

While it would simplify the presentation of cases by ending each one at a definite successful conclusion, most water situations have phases and may never reach the point of being "solved," although that is a good way to state goals. This ongoing nature of water problems can be seen in a scenario like cleaning up an estuary, but even a defined infrastructure project with beginning and ending points is normally followed by assessments and in some cases, substantial revisions and changes.

The cases that follow do not provide an exhaustive set of IWRM situations but they do range across a number of them. The Chesapeake Bay case illustrates an estuary and watershed problem of the type that plagues water resources managers around the world. The Pecos River Compact illustrates a common type of transboundary problem, as does the Virginia Beach pipeline case, which involved a large interbasin transfer. The Missouri River Master Manual illustrates operations within a complex multi-state river basin. The case involving the Chancay–Lambayeque watershed in Peru illustrates an urban-agriculture conflict involving substantial issues of social equity. The Red Fox Meadows stormwater case introduces recreation and environmental learning along with stormwater management. Finally, the Green Mountain Reservoir case illustrates conflicts over water rights in an interbasin transfer setting but with a solution involving a negotiated compensatory storage approach.

Chesapeake Bay

One of the toughest jobs in water resources management is in managing water quality and ecology in estuaries and other coastal waters. Large populations live very near these and the waters support diverse and valuable fish and wildlife species, thus making biological productivity vital despite pressures of human activities. Estuaries are interfaces between river basins and maritime systems, so river basin management and local runoff determine their health.

The United States has many small and large estuaries, and Chesapeake Bay is among the most significant in terms of biological importance and the pressures of development and use. Recognizing threats to it and apparent deterioration, the USEPA and riparian states created the Chesapeake Bay Program in 1983. To the south of Chesapeake Bay are estuaries named Albemarle Sound and Pamlico Sound. This writer was responsible for an early version of part of what became the Albemarle-Pamlico National Estuary Partnership (https://apnep.nc.gov) and will draw on that experience to explain how IWRM works in this context. The writer's participation was as responsible official for a project aimed at restoration of the Chowan River, an important tributary to Albemarle Sound.

The focus of the discussion will be on Chesapeake Bay with examples from the early work on the Chowan River, which has a meaningful scale to explain the IWRM aspects. The Chesapeake Bay program is very large and requires many smaller project efforts like the Chowan. The Chesapeake is the largest bay in the United States, with about 8000 miles of shoreline and 4300 mile^2 of water surface. The drainage basin is about 64 000 mile^2, mainly from five principal rivers. The basic problem is that the Bay has been in decline due to population and overuse with declining seafood harvests, loss of submerged aquatic vegetation, adverse impacts on nursery fish species, and impacts from wastewater.

The situation requiring action for both Chesapeake Bay and the Chowan River was eutrophication, along with impacts visible in algae blooms, fish kills, dead zones, and declining submerged vegetation. Leadership and coordination were evident in both programs, especially in the initial phases. In the case of the Chowan, leadership came in the form of local advocates, and one local leader had a major role in stimulating action. Without that person's leadership, it is possible that the project would have languished. After the initial stimulation provided by local leadership, successful facilitation was required to balance interests and give momentum to the problem-solving process.

The need for technical skills for the Chowan project became evident quickly because the implementation of regulatory controls required scientific proof about the causes of eutrophication to proceed. Without this proof, solution strategies could be challenged with the restoration effort halted. Management skills were required to juggle teams of researchers, public involvement, and the interactions with political and governmental entities. Ultimately, approval for solution strategies required coordination and consensus building among stakeholders and decision bodies, particularly administrative law boards. Funding was also required to provide impetus to the needed research on the water bodies.

The Chowan River seems to have recovered well and there has been success on the Chesapeake Bay Program, although both will face new challenges as the regions continue to develop and their politics change. The political disincentives to collective action may slow momentum and the inter-jurisdictional cooperation needed, particularly in the larger-scale Chesapeake program.

Pecos River Compact

Transboundary river disputes are difficult to resolve in water-short regions and they also affect some humid regions due to timing of shortages or water quality problems. Resolving them requires success in resolving political and legal matters related to allocation of water as well as use of surface and subsurface hydrology. The Pecos River case explains how science and law converged to resolve a longstanding interstate water allocation dispute in the southwestern United States. The case involves US law of interstate compacts, although disputes between independent nations will involve different political and legal mechanisms.

The Pecos River begins in the mountains of north-central New Mexico and flows about 900 miles across semi-arid regions into West Texas and on to its confluence with the Rio Grande River near Langtry, Texas. It drains about $25\,000\,\text{mile}^2$ in New Mexico and $19\,000\,\text{mile}^2$ in Texas.

The dispute dates to when irrigation development began to occur with the Fort Sumner project in 1863, followed by successive periods of other diversions and well pumping. Several dams and reservoirs have been constructed in the basin, including Red Bluff dam as a boundary reservoir at the state line. Controversy about interstate water allocation began early in the twentieth century, and these led to many meetings, studies, and compact commissions. The Pecos River Compact

Commission, which exists today, was formed in 1942 and after several years of negotiations, an interstate Compact was ratified in 1949. The Pecos River Commission has three members, one from each state and a federal commissioner, who lacks a vote.

Unfortunately, the Pecos River Commission had difficulty resolving disputes due to the lack of a tie-breaking vote and this led to a 1973 Supreme Court (SCOTUS) case where Texas as the downstream state sued New Mexico claiming shortfalls in water deliveries. The Court appointed a Special Master, who coordinated a process of dispute resolution and negotiation between the two states. The Special Master reported to the court and recommended remedies for past shortfalls of 340 100 acre-feet for the period 1950–1983, and this was settled by a financial judgment of about $14 million against New Mexico.

The second part of the case was to establish a mechanism to administer the compact in the future, with the agreed-on date of 1947 as the date of equitable apportionment of the water. For enforcement of this arrangement, SCOTUS issued an Amended Decree in 1988 to establish procedures for administration of the Compact and to appoint a River Master to manage them. A River Master's Manual was prepared as a result of settlements made during the litigation.

The situation requiring action was the water shortages in Texas that which occurred due to continuing water development in the upstream state. This is a common problem in transboundary water disputes. Leadership and coordination were provided by the Special Masters appointed by SCOTUS, by a water resources engineering consultant who developed procedures embodied in the River Masters Manual, and by various members of the Pecos River Commission over the years. In disputes of this kind, facilitation to achieve consensus is difficult due to the opposing agendas of the parties. For that reason, negotiation is needed and, in the absence of agreement, a dispute resolution mechanism is required.

The issue required substantial hydrologic, engineering, and legal skill sets. As mentioned, the water resources engineer who developed the mechanism for river management created a unique approach called the "inflow-outflow" method.

Ultimately, the coordination and consensus building processes extended for many years and could only be completed through judicial decisions. Nevertheless, the case illustrates many aspects of IWRM and the reality that some situations require judicial intervention.

Virginia Beach Pipeline

Closely related to transboundary issues of water flowing across state lines are interbasin transfer problems where water is moved from one basis to another but not always from one jurisdiction to another. Such an issue as this involving the Virginia Beach, Virginia, water supply illustrates many interesting aspects of IWRM successes and failures.

The case extends back to the 1970s when Virginia Beach's rapid growth led to a precarious water supply situation. The city had been relying on its neighbor Norfolk for raw water and on limited desalting and wells. Their contract with Norfolk was unfavorable and droughts exposed Virginia Beach's lack of water security. This compelled Virginia Beach to seek an independent water supply.

During the 1980s, Virginia Beach studied many projects ranging from interbasin transfers of water to desalting plants. They also identified other strategies, like demand reduction, regional interconnections, groundwater, and combined strategies. They reduced these to four alternatives, including a plan for an interbasin diversion from Lake Gaston in the Roanoke River Basin. The plan was to take water from an existing reservoir owned by Virginia Power and transport it via a 60-inch pipeline for about 85 miles to the Virginia Beach region.

Virginia Beach needed the cooperation of its sister state of North Carolina to gain permission for the pipeline. The two states had already organized a bi-state committee to discuss alternative solutions to border problems involving river water quality, groundwater withdrawals, and interbasin transfer. The committee met during the 1970s and progress was apparent, but in 1983 a Senate race interfered with politics, and North Carolina announced that it would oppose the project and end the cooperation.

This was followed by a long period of permit applications, litigation, and political challenges. These included environmental assessments and environmental impact statements, district and appellate court rulings, and refusals by the Supreme Court to grant appeals. Every ruling upheld the need for the project and the project was completed in 1997.

The situation of lack of a secure water supply was an existential crisis for the growing city and required action. Leadership emerged in the city government and political leadership and was sustained over many years as Virginia Beach navigated the technical, financial, legal, and political challenges. While facilitation was attempted early in the process, the breakdown ended the cooperation and made facilitation impossible. Technical skills during the process involved engineering, environmental sciences, and legal analysis. The management skills required to balance the many conflicts and decision processes included financial, legal, and engineering, among others. As the mediation became impossible, coordination and consensus building between the parties failed, but within the individual parties such as the city of Virginia Beach, consensus about the project and strategy did emerge.

Missouri River Master Manual

Control of the multi-reservoir system on the Upper Missouri River provides an IWRM case involving issues of operations relating to flood, drought, environmental management, and others. The Missouri River is the longest river in the United States at 2619 miles from its source to St. Louis, Missouri, and it drains diverse ecological regions in 529 350 mile^2 across 10 states and one Canadian province.

The main-stem reservoir system has six dams and reservoirs with a combined system storage of about 73.1 million acre-feet (MAF), which is the largest for any river basin in the United States. These are to be operated for eight purposes: flood control, irrigation, navigation, hydropower, water supply, water quality, recreation, and fish and wildlife enhancement. The problems in the basin include conflicts between upstream and downstream states and municipalities and between alternative uses of the streamflow. The most visible mainstem problem is flood control, particularly in the lower reaches. Drought can cause navigation difficulties, also mainly in the lower reaches. Water supply issues vary by region, from water rights problems in dry headwaters regions to proposals for interbasin transfers from the mainstem. There are conflicts between upper irrigation states that want high water levels and lower states seeking lower levels for flood control. Hydropower generation is below the basin's capacity, mainly due to infrastructure and governance issues. Also, water quality is difficult to characterize on a basinwide basis. Agriculture is a major source of basinwide contamination, especially nutrients that contribute to the dead zone in the Gulf of Mexico. Reservoirs capture sediment, and subsequent erosion and sedimentation cause problems for navigation and infrastructure downstream. On environmental issues, the debate over how the provision of flows and channel depths for navigation has affected the Corps' ability and willingness to meet ecosystem needs.

Governance of the river evolved for irrigation, flood control and eventually to mainstem reservoir construction. This construction was to be followed by a basinwide organization, but it did not succeed. Several attempts at coordination of river basin management have failed as politics block implementation of a mechanism to coordinate decisions and mediate conflicts in the basin and river management remains the responsibility of the USACE, which must weather political controversies and legal challenges.

The USACE filled the gap due to lack of organization and beginning in 1960 published a Master Manual as the river coordinating mechanism that endures to this day. The Master Manual has been revised a few times, to include adding environmental responses and adaptive management tools, but each revision is controversial and some draw legal challenges. No matter how the USACE manages the water, someone is unhappy and political leaders blame government officials in order to shift public attention away from themselves.

The manual is expected to balance legal, political, and management issues in day-by-day river management. River politics involves the diverse agendas of elected officials, and legal constraints stem from legislation, water rights and lawsuits. The resulting straitjacket binds the USACE in a tight balancing act to do the best that is can despite the inevitability of dissatisfaction.

The situation requiring action in this case is to provide a multi-state coordinating mechanism for the Missouri River, which serves many demands. Without such a mechanism, floods, droughts, and other water-centric problems would spiral out of control. As the states failed continually to find a solution to the need for joint management, the USACE has provided the leadership and coordination. Facilitation toward agreed actions takes place continually through USACE public involvement processes, but some parties express dissatisfaction despite these efforts. Technical skills to assess conditions and make good control decisions are essential, and the management of the infrastructure, the decision-making processes, and the public involvement requires high levels of skills as well. While coordination and consensus building are desirable, and significant strides are made, the record thus far indicates that a certain degree of disharmony is inevitable in this large-scale and diverse basin.

Chancay–Lambayeque Watershed in Peru

The Chancay–Lambayeque basin is one of Peru's most important basins due to its size and multiple uses in a semi-arid area (Global Water Partnership 2022). Agriculture is the main economic activity in the basin and uses most of the available water. Equity is a critical issue, and a majority of the population lacks access to safe and reliable drinking water. Water scarcity is evident in malnourishment and water-related disease, conflicts among users, and degradation of ecosystems along with salinization and soil erosion.

In the lower part of the watershed the San José/Pampa de Perros' farmer community irrigates some 98 000 ha of mostly rice and sugar cane. Scarce water supplies force farmers to irrigate with poorly treated or untreated wastewater below Chiclayo city and this creates health issues, including a cholera outbreak in 1983. As a solution, the regional authorities proposed direct discharge of the wastewater of the city to the sea and to bypass the San José farmer community. This would effectively destroy the economy of the farmers, and protests broke out when farmers obstructed the drain into the sea and caused floods on city streets. The regional authorities had proposed to move the farmers and use wastewater for forestry, but the farmers wanted to remain and use treated waters for agriculture.

After initial social action, a Technical Commission was formed among representatives of the Chiclayo Municipality, regional authorities, and CES Solidaridad. A citizen group, the CES Solidaridad to search for a solution. It proposed a project for the "Future Development of San José Farmer Community Wastewater." The project would include wastewater treatment in the Master Plan of Chiclayo with funding from the German development bank KFW. A technical team to work on the project included Solidaridad and professionals from the local University.

The proposed solution was to use wastewater treatment lagoons, and KFW asked the community to give up 146 ha of land to build 18 of them. After negotiations between the community, land owners, and the Municipality, agreement was reached such that treated waters would be used for agricultural production with control and distribution by the community and support from CES Solidaridad.

The treatment ponds were constructed from funds from a Cooperation Agency (Pan para el Mundo), CES Solidaridad, the Region, and KFW. The farmers have now demonstrated the potential of using treated wastewater to irrigate the sandy pampas areas, and this has helped overcome water scarcity and add value to the land as well as to alleviate poverty.

The situation requiring action was the deteriorating water quality and health issues among the farmer community, along with the proposed solution to simply relocate it. Cases like this with significant social action usually involve leadership and coordination to initiate dialogue and hold the stakeholders together. The negotiating group involving diverse actors required successful facilitation, which can often be provided by an outside non-involved party, like the German development agency staff. Technical skills required in this case were available from the local university, and management skills were augmented by KFW as they negotiated with the city to provide land for the project. The coordination and consensus building in this case seemed successful, despite the disparate goals of the parties.

Red Fox Meadows Stormwater Project

The Red Fox Meadows Natural Area in Fort Collins, Colorado, is a combined urban wildlife refuge and stormwater detention area that is managed jointly by the city's Utilities and Natural Resources Departments (Fort Collins Utilities 2022). Benefits include recreational and educational opportunities, wildlife habitat, stormwater pollution prevention, and flood protection. The 40-acre facility has about 80 wildlife species and 126 plant species.

The project exhibits several common attributes that require an IWRM approach in cities. A problem of increasing stormwater threats due to urbanization and climate change demanded action, and rather than install a large pipe and then devote the site to new housing, the city conceived the new plan to provide multiple benefits. Developing plans like this requires good cooperation and visioning among staff in different departments and on appointed or elected citizen boards.

The leadership and coordination required for a project of this type often come from champions within the staffs of cities. In many cases, cities employ visionary professionals in departments like engineering, natural resources, and planning. It may take perseverance for this kind of leadership to emerge, but it can result in innovative solutions with far-reaching effects. Successful facilitation to nurture a project requires support from leaders with enough authority to command attention and liaise with higher level managers and elected officials. Technical and management skills are required, and when cities lack such capacity, consultants may bridge the gaps and provide the help. Coordination and consensus building are particularly important here because the long term nature of the planning and development process offers many opportunities to sidetrack or derail the project.

The results of the project are improved flood protection, better water quality and increased wildlife habitat in Red Fox Meadows for future visitors. Elementary school students are frequent visitors to the outdoor classrooms for nature studies. They identify plants and animals, study food chains and webs, learn ecological concepts, and survey plants and animals. They also learn how stormwater runoff from lawns and streets contains pollutants and how detention areas provide treatment by settling out these pollutants before the water enters streams and ponds.

Green Mountain Reservoir

The Green Mountain Reservoir case illustrates attributes of IWRM that occurred decades in the past. The reservoir stores Blue River water in Summit County, Colorado, and represents a compromise that has benefits for all parties. Green Mountain Dam was built during the Franklin Roosevelt presidency between 1938 and 1942 by the USBR. The reservoir was built to benefit Colorado's Western Slope, which stood to lose trans-basin water as a result of the Colorado–Big Thompson Project (C–BT project). The C–BT project provided supplemental irrigation water to enable agricultural development in northern Colorado's over stressed South Platte River basin.

The story unfolded during the Great Depression as farm interests in northern Colorado recognized that unallocated water on the west slope could be exploited to benefit the East. Western Colorado interests saw this as a threat to their long-term future and stood ready to block any initiatives. With slope water had already been diverted to benefit the Denver area, and west slope leaders worried that further losses of water would threaten the future.

The situation requiring action was the need for supplemental water in northern Colorado without threatening Western Colorado. Leadership and coordination involve negotiation among the two interest groups and involving federal and state governments, so local leaders as well as appointed and elected officials needed to take leadership jointly. The facilitation required was to keep negotiations going in a productive way so as to achieve some consensus. The USBR was able to provide substantial technical skills and management skills as well to create a complex project to meet all needs. Ultimately, this was a successful example of IWRM in action, although any long-term diversion of interbasin waters creates unforeseen effects in growth, environmental chain, and social impacts.

Analysis and Discussion

Of the six required requisites for successful IWRM, technical skills and management skills are required at all. To analyze the other for situations, the remaining attributes can be compared. Table 23.1 compares them.

The situations show substantial variation in problems, leadership, facilitation, and consensus building. They illustrate water management purposes that include water supply, wastewater, stormwater, irrigation, and flood control. Environmental water issues loom large in several of the cases. Leadership interventions ranged from local leaders to city staff to political leaders at different levels and to

Table 23.1 Comparison of IWRM case studies.

	Situation requiring action	Leadership provided	Facilitation provided	Consensus-building
Chesapeake Bay	Estuary eutrophication	Local advocates	Governmental or NGO	Administrative law boards
Pecos River Compact	Water shortages for agriculture	Special masters	Special masters	Judicial
Virginia Beach pipeline	Water shortages for urban development	Government and political leadership	Did not work	Judicial
Missouri River master manual	Many conflicting water management demands	USACE	USACE	USACE
Chancay–Lambayeque watershed in Peru	Water quality and health	Local and NGO leaders	NGO	NGO
Red Fox Meadows stormwater project	Stormwater threats	City staff	City staff	Public involvement via city processes
Green Mountain Reservoir	Water shortage, threated water losses	Local leaders and USBR	Political and appointed leaders	Political and appointed leaders

government appointed leaders. Facilitation was similar, but the Peru case illustrates how an NGO can provide useful intervention. Consensus building is often difficult and in three of the seven cases it required use of the judicial system. The larger scale problems tend to toward judicial solutions more than smaller scale problems. This indicates again the importance of the scale factor in water resources planning where people who can talk to each other directly have a better chance than those who work through representatives to negotiate consensus solutions.

Ultimately, the complexity and variation among the cases illustrate why it is difficult to explain IWRM as a definite process to be applied. The water resources manager must use tools and knowledge available from training and experience and apply them toward each individual situation with its unique scenario, setting, and players. This is also evident in the many cases posted on the Global Water Partnership site.

References

Fort Collins Utilities (2022). Red fox meadows natural area restoration. https://www.fcgov.com/utilities/what-we-do/stormwater/drainage-improvement-projects/canal-importation-ponds-and-outfall/red-fox-meadows-natural-area-restoration (accessed 5 May 2022).

Global Water Partnership (2022). Peru: Treated waters – communal participatory management and its impact on human development and ecosystem (#436). https://www.gwp.org/en/learn/KNOWLEDGE_RESOURCES/Case_Studies/Americas-Caribbean/Peru-Treated-waters-communal-participatory-management-and-its-impact-on-human-development-and-ecosystem-436 (accessed 5 May 2022).

Questions

1 Why do words with broad meanings like "management" and "integrate" lead to different ways to understand water resources management?

2 What are the differences in the meanings of the words leadership, coordination, facilitation, and consensus-building? Why are they so important in IWRM situations?

3 What three levels of water management were outlined in the chapter? Name scenarios to illustrate them?

4 Do the tools and methods of IWRM apply to both developed and developing country situations? Why or why not?

5 What attributes distinguish IWRM from other levels of water management?

6 Can a single blueprint for IWRM be presented? Why or why not?

7 What is meant by the context of water issues and how do it relate the scenario?

8 What is meant by use of the word "reality" in the following quote from the Global Water Partnership: "... it is necessary to select the group of instruments that better suit a specific reality, considering the existing social and political consensus, available resources, and geographical, social, and economic contexts."

9 Explain your concept of what comprises a "problem" to be addressed by IWRM.

24

Careers in Water Resources Management

Introduction

As the previous chapters explained, the issues you can face in water resources management provide important and interesting challenges. This chapter outlines the types of jobs and employers that can provide you with gateways into this productive sector of work. Many types of jobs are available in organizations devoted to water management, water uses, water industry equipment and services, policy and planning, and regulatory work. Some of the jobs may seem basic, but all work in water management contributes to stewardship of water. Their common thread is managing wet water and the facilities needed for all water management purposes.

If you work in the water industry, your job may be in the private sector, but you will work a lot with the government at all three levels: local, state, and national.

The water industry is like a constellation of different organizations and is distributed across many types of employers and jobs. It is a big industry, but not like a giant corporation with a central headquarters and a lot of satellite operations or a government chain of command reaching from the top to the bottom. How it is organized into parts was explained in Chapter 12. It has water handlers like drinking water utilities or government agencies that own and operate dams; water users like industries or electric power utilities; suppliers of equipment and services; and government agencies for planning and regulation.

Water Industry Employers

This table provides a list of types of employers with approximate numbers of jobs.

Water Resources Management: Principles, Methods, and Tools, First Edition. Neil Grigg.
© 2023 John Wiley & Sons, Inc. Published 2023 by John Wiley & Sons, Inc.

Water supply utilities	Provide drinking water to homes and businesses. About 50 000 utilities or community water systems employ some 250 000 workers.
Wastewater utilities	Collect used water from homes or businesses. About 225 000 jobs people across some 30 000 organizations, mainly as city departments or utilities.
Stormwater and flood control organizations	Provide management and control of runoff. About 50 000–100 000 jobs mostly in local governments. Green infrastructure may increase jobs.
Irrigation and water management districts	Provide water for farming and landscaping. Mostly in the western US but some in humid areas such as Florida. Job count uncertain because sector involves diverse collection of organizations.
Government water managers	About 50 000 jobs in all federal water agencies. The Corps of Engineers has the most jobs. State governments have some water industry jobs, mainly as regulator.
Electric power utilities	Many have jobs in water management for hydropower and thermoelectric water use.
Water users	Large water users have water related jobs in multi-family residential, commercial, industrial, food processing, landscaping, and other sectors.
Professional services	The main type of organization is consulting engineering, the largest sector of employment devoted to work in the water industry.
Constructors and service providers	Construction companies and related activities such as well-drilling are a source of jobs related to the water industry.
Equipment and materials vendors	Provide equipment such as pumps, large pipes and valves, and controls equipment.
Other water industry support groups	Associations, advocacy groups, research organizations, and financial industry.

Water Industry Work

Water management work focuses on engineering and other professional roles, and pathways to them are through education and experience. Starting with jobs like system operator, monitoring, or maintenance can lead into more advanced work. Some examples of types of work follow.

Developing Master Infrastructure Development Plans

Consulting civil and environmental engineers develop master plans for water infrastructure systems. These plans outline how a service, such as a drinking water system, will be expanded and improved in the future.

Planning Infrastructure Projects

Planning water infrastructure facilities is frequently assigned to civil and environmental engineers. At the larger scale, they might involve a new reservoir and/or the associated water handling facilities. This involves tasks like environmental permits, lining up finance, and obtaining public support, as well as the technical aspects of the plans.

Performing Asset Management Studies

Aging infrastructure is a tremendous challenge for water and wastewater utilities. Managers in water and wastewater utilities organize activities from inspection of pipelines to statistical studies to determine strategies to renew them.

Planning Utility Rate Increases

Rate increases are common scenarios for wastewater utilities. The analysis can involve a range of disciplines.

Wastewater Plant Operation and Recycling

Wastewater plants have been in operation for more than 40 years, and now they are being renewed and upgraded. The work involves mainly engineers, operators, and laboratory personnel, but life scientists, economists, and other disciplines now help convert them into Utilities of the Future.

Water Recycling Systems

Many projects for water recycling will be implemented in the future because of the importance of making water systems more efficient. The work involves engineering and science, along with economic rate studies.

Planning Green Infrastructure

The implementation of green infrastructure is an emerging trend in the water industry. Other transformations in cities may occur as these are implemented and water managers will work closely with urban planners.

Stormwater Utility Management

The challenge to finance storm water infrastructure has led to organization of utilities where funding comes from assessment of fees rather than from general taxes. This work involves watershed analysis, engineering planning and assessment, and financial studies.

Operating Multipurpose Reservoirs

The operation of these reservoirs requires careful decisions by the operators who must consider the multiple functions without upsetting constituents and violating guidelines.

Minimizing Flood Damages

This involves floodplain planning and management to evaluate the best way to minimize the flood damages in a metropolitan area. To prepare the maps required for this program requires planning and engineering work with typical functions to map floodplains, assess vulnerability of properties and people, and to determining the best strategies to minimize flood damages.

Permitting

The manufacturing and energy industries using water need permits for their water operations. Work required to prepare environmental studies and apply for permits involves planning, engineering, data collection and ecological studies.

Estuary Recovery

Regulators face pollution problems with complex interactions among different sources of pollution. Effects may be difficult to evaluate on a scientific basis and require research. Managers can normally not mandate solutions and voluntary actions are needed.

Assessing Failures and Remedial Actions

Probing failures in drinking water, water quality, or buried infrastructure involves complex issues and requires detailed studies. An important challenge is to control the rate of failures in buried infrastructure like distribution pipelines.

Water Industry Jobs

It takes multiple players with different roles to address issues like those listed in the previous section. They have different job titles, which water managers can aspire to. These are listed and explained briefly, with selected examples given.

Executives and Managers of Water Agencies and Utilities

Managers in the water industry have responsibility for all operations of their organizations. Many of them started as engineers, but other career routes to management are available. They must juggle technical and non-technical tasks and they interact extensively with the public and elected officials. With so many water organizations, executive and managerial jobs are available in many places. Many organizations are small, but even running a small water management agency offers many interesting challenges. An example of a route to upper management might be an engineer who displayed skill in infrastructure management. The engineer would learn new skills on the way up the promotion ladder and be a seasoned management on arrival at the top.

Civil and Environmental Engineers

Civil and environmental engineers build and manage water infrastructure facilities and processes. While their focus is infrastructure, they are often assigned management tasks. A civil or environmental engineer working in the private sector might establish a reputation as an innovative thinker and soon become an advisor to water utilities. This could lead to an appointment as a lead engineer in a major consulting firm and to leadership roles in national associations.

Water Resource Scientists, Including Hydrologists

The main water resources scientists are hydrologists and ecologists. These are essential because many water organizations obtain and manage their own water resources, which involves watershed assessment and monitoring. Hydrologists and ecologists work in research, field studies, and laboratory analysis. Ecologists are needed to study water habitat and help with environmental assessments. A hydrologist might graduate in watershed sciences and find a post in a stormwater organization as responsible for the floodplain management program. This can lead to opportunities in public speaking and work with association committees, which leads to national recognition in the field of flood plain management.

Watershed Managers

A watershed manager performs interdisciplinary work to address science and management issues. Watershed science explains the functioning of watersheds and watershed management builds on it to determine needed responses to drivers of change in watersheds. Examples might be how to modify water flows or improve water quality. Jobs as watershed scientist and watershed manager often occur with different names, such as wetland scientist, water resources engineer, environmental planner, hydrologist, watershed ecologist, and water quality analyst. A watershed manager might take over responsibility to manage a city's water supply watershed for purpose of keeping source water clean. This role might evolve into a management post with exposure to policy and political issues.

Water System Operators

The water industry involves many technical systems like treatment plants, pipeline systems, or dams. They require a technical field of operations and job of operator. These jobs require training in systems operation, and you can enter the field from different backgrounds. The jobs range in responsibility from systems in large metropolitan areas to small systems serving only a few 100 people. An operator of a water treatment plant might confront complex problems, like chemical scaling. These require expertise in chemistry and troubleshooting. The operator might build on a background in chemistry and biology for the task, and this work might lead to promotion into management.

Construction and Maintenance Managers

Managing the infrastructure of the water industry requires construction specialists and administrators to deal with new and aging systems. Construction managers in utilities are involved with project management for new and renewed facilities from concept to completion. A construction manager might have to assess failure causes and recommend ways to renew systems. Starting with the field of construction might lead into an expanded role in asset management.

Management Consultants

Many jobs that support the water industry are in consulting firms. Their staffs comprise a shadow workforce for water organizations and their workers often have similar skills as utility personnel for planning, design, and management. After working in another field, a consultant might obtain advanced training in civil engineering and because an expert in finance of water systems. This can lead to establishment of a company to advise water utilities on their rate structures.

Physical and Life Scientists

Chemistry and biology are important sciences for water management work, and they have subdivisions, such as microbiology. Chemists and biologists may work in laboratories and perform diagnostic tasks in support of overall management.

This can lead to promotion to laboratory manager and to upper management in a utility.

System and Process Engineers

Water infrastructure systems involve complex infrastructure and automatic controls that require electrical, mechanical, and chemical engineers. One of them could start in an electric power utility and gravitate toward controls work for environmental systems. This can lead to work with permits and other regulatory matters and to promotion within the electric power utility.

Information Technology Specialists

Data bases, models, and control systems require specialists in information technology. They can enter the water field from mathematics, computer science, statistics, and fields of engineering. An IT specialist might become involved with modeling of water systems and work for a water utility. This could provide ideas about new products and to work as a software developer with contacts across the water industry.

Landscape, Irrigation, and Water Conservation Specialists

With water being scarce and with rising demand set to collide with environmental needs, water conservation will grow in importance and require specialists in landscape, irrigation, and water use studies. An irrigation specialist in production agriculture might find a job with a water utility seeking to encourage customers to transform landscaping to water-efficient patterns. The specialist might meet with community groups to explain the program and become involved as a representative of the utility to the city.

Public Information and Education Specialists

Work in the water industry involves the public in many ways and requires public meetings, brochures, and news releases to inform the public about water. Broad backgrounds in journalism, education, communications, and related fields are helpful for these positions. A communications specialist with a degree in history might work in outreach for a water management organization. Soon the specialist will learn about the organization's work and represent it to the public. This work can be central to informing the public and to the organization's success.

Managers of Financial, Administrative, and Legal Affairs

Water organizations face complex issues of public finance, labor and personnel rules, and compliance with regulatory requirements. The large investments in infrastructure require specialists in contracts, and the workforce requires personnel officers. These tasks involve business administration, human resources fields, or law. A new law graduate can work within a large water organization and rise to executive status.

Technical Sales and Marketing

Successful infrastructure systems require many kinds of equipment and supplies. The convention expositions of water associations will feature many booths to display various products and services. Representatives of the organizations will vary from their chief executives to sales people, and include highly competent technical professionals. A utility engineer might become a supervisor of a water distribution system and take interest in the pipe, valves, controls, and other components needed. The engineer may find opportunities in representing the pipe industry on the staff of a national association.

Planners and Policy Specialists

As water management is by all three levels of government and many stakeholders are involved, planning and policy analysis occur on a widespread basis and involve everything from a small water district planning its water conservation program to Congress debating a proposed new statute. A history major might be attracted to public administration and work in a state government post. This might lead to opportunities in a state government department of water resources and to involvement in policy issues.

Career Paths in the Water Industry

You can prepare for work in the water industry through academic or vocational routes. Upward mobility can result from getting your foot in the door in an entry-level job and proving yourself by work and learning. Academic fields to prepare for the water industry are aligned with jobs like managers, engineers, scientists, administrators, lawyers, and support workers. Many are pathways to similar jobs in related sectors such as electric power, energy, and telecommunications. The water industry is somewhat distinctive in the way it addresses issues of health and environment. Vocational paths to water industry jobs might be attractive to people with military or industry training, operator licenses and construction management.

How you prepare for work in the water industry depends on which position(s) you might like to pursue. When utilities and other organizations recruit for these positions, they will look for certain skills and preparation. While there are some exceptions, executives and managers usually work up through water industry organizations rather than to be hired from another industry. Probably the reason is the in-depth knowledge needed to comply with laws and to deal with complexities of managing water that are not always evident to outsiders, such as the politics and public opinions about water management.

No matter the job, studying engineering is a common path into the water industry. Civil engineering is a good choice for water and wastewater utilities. Information is available from the American Society of Civil Engineers (ASCE) at www.asce.org. Lists of accredited schools can be found at the web site of the Accreditation Board for Engineering and Technology, at www.abet.org. In addition to civil engineering, you can find information about related fields, especially environmental engineering. Many engineers start work in consulting companies and transition to local governments, including utilities.

Engineering technology as a field for preparation is oriented toward practical applications, which may appeal to students with an affinity for technology but have chosen not to study engineering. The programs might be two-year or four-year. The US Bureau of Labor Statistics wrote: "Engineering technicians usually begin by performing routine duties under the close supervision of an experienced technician, technologist, engineer, or scientist. As they gain experience, they are given more difficult assignments with only general supervision. Some engineering technicians eventually become supervisors."

Technician training is good preparation to become a treatment operator, and training can also be obtained at a community college. Distribution and collection systems operators tend to come from construction and maintenance occupations, rather than chemistry and hydraulics. For these jobs you need aptitude and experience in construction and equipment trades. Taking an entry position and then becoming trained and certified may be a good route, and state associations and agencies may offer continuing education. A local community college may offer a specific course of instruction for water and wastewater utility operators. The programs can range from non-credit courses of a few hours to programs that lead to a two-year Associate Degree. The result may be a degree or a non-degree certificate. Community college, technical college, and trade school programs can be found from Internet searches.

The work of environmental and conservation occupations in water and wastewater utilities is diverse and possibilities for training include the life sciences (agricultural, biological, conservation, or forest sciences); physical sciences (atmospheric, chemical, environmental, hydrologic, or geosciences); and social sciences (economics and urban and regional planning).

As a rapidly changing field, IT positions in utilities are evolving. People who can service computers, handle software and operating systems, and operate information networks can qualify for these positions. Previous experience with special IT applications will be helpful, including SCADA, facility mapping, enterprise work management, data bases, and simulation models.

Another career path is to start in a laboratory. Employees in water and wastewater labs might have degrees in chemistry or microbiology and preparation can be done at colleges and universities.

Appendix A

Units, Conversion Factors, and Water Properties

Introduction

This appendix provides examples of basic units in English and SI systems, calculations, conversion factors, and selected data. The examples are simple because the goal is to illustrate basic approaches to water resources management.

Water Management Units in English and SI Systems

The below table shows common water management units and abbreviations at small, medium, and large scales for volumes and rates. Units for lengths, areas, and weights are the same as for other applications, feet and meters, acres and hectares, pounds and kilograms, etc. Units for power and energy are also the same, horsepower, kilowatts, and kilowatt-hours.

For drinking water, consumption is usually in gallons per capita per day (gpcd). For water tanks or small ponds the contents might be in gallons, thousand gallons, million gallons, or cubic feet. For reservoirs the contents might be in million gallons, cubic feet, or acre-feet. SI units might be MCM. Major bodies might have volumes in million acre-feet or BCM. Rates of flow for household water tap, pump, or small well might be given in gpm or l/s. A pipeline will normally be in gpm, MGD, or cfs. A treatment plant is usually in MGD. Streamflows are normally in cfs. Comparable SI units are shown in the table.

Water Resources Management: Principles, Methods, and Tools, First Edition. Neil Grigg.
© 2023 John Wiley & Sons, Inc. Published 2023 by John Wiley & Sons, Inc.

	Application scale	English	SI
Volume	Drinking water	gal, ft³, CCF, TG	l, CM
	Tank contents	TG, MG	MCM
	Stream yield, reservoir	AF, cfs-day, TAF, MAF	MCM
	Major scale	mi³	km³, BCM
Rate	Household, small pump	gpm	l/s
	City supply	MGD	MCM/day
	River discharge	ft³/s	m³/s

Abbreviations

AF	Acre-feet
BCM	Billion cubic meters
CCF	Hundred cubic feet
cfs-day	cfs for a day
ft³	Cubic feet
ft³/s	Cubic feet per second
gal	Gallons
gpm	Gallons per minute
km³	Cubic kilometer
l	Liter
m³	Cubic meter
m³/s	Cubic meters per second
MAF	Million acre-feet
MCM	Million cubic meters
MCM/day	Million cubic meters per day
MG	Million gallons
MGD	Million gallons/day
mi³	Cubic mile
TAF	Thousand acre-feet
TG	Thousand gallons
l/s	Liters per second

Conversion Factors

	Volume			
1	acre-foot (AF)	=	43 560	cubic feet (ft^3)
1	acre-foot	=	325 853	gallons (gal)
1	acre-foot	=	1 233.49	cubic meters (m^3)
1	thousand AF (TAF)	=	1.2335	million m^3 (MCM)
1	million AF (MAF)	=	1 233.5	million m^3 (MCM)
1	cubic foot	=	7.4806	gallons
1	cubic foot	=	28.317	liter
1	gallon	=	3.7854	liter
1	million gal (MG)	=	3.0689	AF
1	billion gal (BG)	=	3.7855	MCM
1	cubic meter (m^3)	=	35.3145	ft^3
1	cubic meter (m^3)	=	1 000	liters
1	cubic meter (m^3)	=	264.17	gallons
1	million m^3	=	810.7	AF
1	billion m^3	=	810 684	AF
1	billion m^3	=	1	km^3
1	cfs-year	=	723.97	AF
1	cfs-day	=	1.9835	AF
1	mi^3	=	3.3792	MAF
1	mi^3	=	4.1682	BCM
1	thousand gal	=	1.337	Hundred cubic feet (CCF)

	Rate			
1	cfs	=	448.836	gpm (gal/min)
1	cfs	=	28.317	liter/s
1	cfs	=	723.97	AF/year
1	cfs	=	1.9835	AF/day
1	cfs	=	2446.6	m^3/day
1	mgd (Mgal/day)	=	1.5472	cfs

(Continued)

	Rate			
1	mgd	=	1120.1	AF/year
1	m³/s	=	35.3145	cfs
1	m³/s	=	22.82	mgd
1	MCM/day	=	408.7	cfs

	Weight			
1	lb		0.453 59	kg

	Area			
1	acre	=	43 560	ft²
1	acre	=	0.404 69	hectare (ha)
1	square mi	=	640	acres
1	square mi	=	259	hectare (ha)
1	hectare (ha)	=	10 000	m³

	Length			
1	foot	=	0.3048	meter
1	meter	=	39.37	inches
1	meter	=	3.2808	feet
1	inch	=	25.4	mm
1	mile	=	1 609.35	meters
1	kilometer	=	0.621 37	miles

Basic Water Properties

In the United States, the liquid gallon is legally defined as exactly 231 inches³.

Temp, °F	Specific gravity	Specific weight, lb/ft³	Specific weight, lb/ft oz	Specific weight, lb/pint
32.2	0.9999	62.42	0.0652	1.0430
39.2	1.0000	62.427	0.0652	1.0432
50	0.9997	62.41	0.0652	1.0429

(Continued)

Temp, °F	Specific gravity	Specific weight, lb/ft^3	Specific weight, lb/ft oz	Specific weight, lb/pint
60	0.9990	62.37	0.0651	1.0422
70	0.9980	62.30	0.0651	1.0410
80	0.9966	62.22	0.0650	1.0397
100	0.9931	62.00	0.0648	1.0360
200	0.9630	60.12	0.0628	1.0046

Sea water is heavier, about 64 lb/feet^3.

Note: The specific weight of water is used often in power and energy computations and the value of 62.4 lb/feet^3 is usually assumed. Corrections for temperatures can be made if needed.

Appendix B

Acronyms and Abbreviations

404	Dredge and fill program of CWA
7Q10	7-day, 10-year low flows
AFA	American Fisheries Association
AMSA	Association of Metropolitan Sewerage Agencies
AMWA	Association of Metropolitan Water Agencies
APA	Administrative Procedures Act
APWA	American Public Works Association
ARS	Agricultural Research Service
ASCE	American Society of Civil Engineers
ASDSO	Association of State Dam Safety Officials
ASDWA	Association of State Drinking Water Administrators
ASFPM	Association of State Floodplain Managers
AWWA	American Water Works Association
AwwaRF	AWWA Research Foundation (now WRF)
BCA	Benefit–cost analysis
BEA	Bureau of Economic Analysis
BIA	Bureau of Indian Affairs
BLM	Bureau of Land Management
BLS	Bureau of Labor Statistics
BMP	Best management practice
Burec	Bureau of Reclamation (also USBR)
CAP	Customer assistance program
CAPEX	Capital expenses
CDC	Centers for Disease Control
CEQ	Council on Environmental Quality
CFR	Code of Federal Regulations
CMMS	Computerized maintenance management system
CMOM	Capacity, management, operation, and maintenance

Water Resources Management: Principles, Methods, and Tools, First Edition. Neil Grigg.
© 2023 John Wiley & Sons, Inc. Published 2023 by John Wiley & Sons, Inc.

COE	US Army Corps of Engineers (also USACE)
CRS	Congressional Research Service
CSO	Combined sewer overflow
CWA	Clean Water Act
DBP	Disinfection byproduct
DHS	Department of Homeland Security
DO	Dissolved oxygen
DOC	Department of Commerce
DOD	Department of Defense
DOE	Department of Energy
DOI	Department of the Interior
DOT	Department of Transportation
DPSIR	Driver-Pressure-State-Impact-Response
DSS	Decision support system
EIA	Environmental Impact Analysis
EIS	Environmental Impact Statement
EPA	Environmental Protection Agency (also USEPA)
EPRI	Electric Power Research Institute
ERS	Economic Research Service
ESA	Endangered Species Act
ET_c	Crop evapotranspiration
FASB	Financial Accounting Standards Board
FBI	Federal Bureau of Investigation
FEMA	Federal Emergency Management Agency
FERC	Federal Energy Regulatory Commission
FHWA	Federal Highway Administration
FIA	Flood Insurance Act or Federal Insurance Administration
FONSI	Finding of No Significant Impact
FPA	Federal Power Act
FS	Forest Service (also USFS)
FWS	Fish and Wildlife Service
GAAP	Generally Accepted Accounting Practices
GAO	Government Accountability Office
GASB	Government Accounting Standards Board
GIS	Geographic Information System
GO bond	General obligation bond
GWP	Global Water Partnership
H&H	Hydrology and hydraulics
HAZUS	FEMA multi-hazard loss estimation software
HEC	Hydrologic Engineering Center
HHS	Health and Human Services

HUC	Hydrologic Unit Code
HUD	Housing and Urban Development
I&I	Infiltration and inflow
IBT	Inter-basin transfers
ICID	International Commission on Irrigation and Drainage
ICMA	International City/County Management Association
ICS	Incident Command System ICS
IDF	Intensity–Duration–Frequency curves
IRP	Integrated resource planning
ISAC	Information Sharing and Analysis Center
ISF	Instream flows
ISO	International Standards Organization
IT	Information technology
IWA	International Water Association
IWMI	International Water Management Institute
IWRM	Integrated water resources management
IWRM	Integrated Water Resources Management
IWR-MAIN	Institute of Water Resources, Municipal and Industrial
JMP	Joint Monitoring Programme
LCR	Lead and Copper Rule LCR
LID	Low impact development
M&I	Municipal and industrial
MCDA	Multi-criteria decision analysis
MCL	Maximum contaminant level
MHFD	Mile High Flood District
MMS	Maintenance management system
MS4	Municipal Separate Stormwater Systems program
NACWA	National Association of Clean Water Agencies
NAFSMA	National Association of Flood and Stormwater Management Agencies
NARUC	National Association of Regulatory Utility Commissioners
NASA	National Aeronautics and Space Administration
NASQAN	National Stream Quality Accounting Network
NAWC	National Association of Water Companies
NAWQA	National Water Quality Assessment Program
NCAR	National Center for Atmospheric Research
NCDC	National Climatic Data Center
NEPA	National Environmental Policy Act
NFIP	National Flood Insurance Program
NGO	Non-governmental organization
NIMS	National Incident Management System NIMS

NOAA	National Oceanic and Atmospheric Administration
NPDES	National Pollutant Discharge Elimination System
NPDWR	National Primary Drinking Water Regulations
NPS	National Park Service
NRCS	Natural Resources Conservation Service
NWIS	National Water Information System
NWRA	National Water Resources Association
NWS	National Weather Service
O&M	Operations and maintenance
OMB	Office of Management and Budget
OPEX	Operating expenses
PAHO	Pan American Health Organization
PCB	Polychlorinated biphenol
PDSI	Palmer Drought Severity Index
PHS	Public Health Service (also USPHS)
PI	Performance indicator
PILOT	Payment in lieu of taxes
PLC	Programmable logic controller
PMF	Probable Maximum Flood
PMP	Probable Maximum Precipitation
POTW	Publicly owned treatment works
PPBS	Planning-programming-budgeting system
PSC	Public service commission
PUC	Public utility commissions
PVO	Private volunteer organization
RCRA	Resource Conservation and Recovery Act
ROD	Record of Decision
ROR	Rate-of-return
SBA	Small Business Administration
SCADA	Supervisory control and data acquisition
SCOTUS	Supreme Court of the United States
SDG	Sustainable Development Goal
SDWA	Safe Drinking Water Act
SEA	Strategic Environmental Assessment
SEC	Securities and Exchange Commission
SFWMD	South Florida Water Management District
SIA	Social Impact Analysis
SOC	Standard Occupational Classification
SPF	Standard Project Flood
STORET	STOrage and RETrieval (EPA database)
SWAT	Soil and Water Assessment Tool

SWTP	Stormwater treatment plant
TDS	Total dissolved solids
THM	Trihalomethane
TMDL	Total maximum daily load
TTT	Tariffs, taxes, and transfers
TVA	Tennessee Valley Authority
UNEP	United Nations Environment Programme
UNESCO	United Nations Educational, Scientific, and Cultural Organization
UNICEF	United Nations Children's Fund
USACE	US Army Corps of Engineers
USBR	US Bureau of Reclamation
USC	The United States Code USC
USCID	US Committee on Irrigation and Drainage
USCOLD	US Committee on Large Dams
USDA	Department of Agriculture
USGS	US Geological Survey
WAPA	Western Area Power Administration
WEAP	Water Evaluation and Planning System
WEF	Water Environment Federation
WHO	World Health Organization
WOTUS	Waters of the United States
WRF	Water Research Foundation
WRPA	Water Resources Planning Act
WTP	Water treatment plant
WWTP	Wastewater treatment plant

Appendix C

Associations, Federal Agencies, and Other Stakeholders of the Water Industry

Associations and Interest Groups

Regulatory Organizations

Association of State Drinking Water Administrators
Association of State Dam Safety Officials
Interstate Conference on Water Problems
Association of State Flood Plain Managers
Association of Clean Water Administrators

Elected Officials

National Governor's Association
National League of Cities

Civil Society Groups

League of Women Voters

Water Provider and Wastewater Agencies

American Water Works Association
Water Environment Federation
National Association of Clean Water Agencies
Association of Metropolitan Water Agencies
National Association of Water Companies
American Public Works Association
Rural Water Associations

Water Resources Management: Principles, Methods, and Tools, First Edition. Neil Grigg.
© 2023 John Wiley & Sons, Inc. Published 2023 by John Wiley & Sons, Inc.

Hydropower, Navigation, Flood Control Interests

National Water Resources Association
  US National Committee for Irrigation and Drainage
  National Association of Flood and Stormwater Management Agencies
  National Water Congress
  State Water Congresses

Agriculture and Resource Development Sectors

Water User's Associations
  Landscape Interests
  Farm Bureau
  Cattlemen's Associations
  Irrigators Associations
  Association of Soil Conservation Districts
  Grain and Feed Association
  Cattle Feeder's Association

Industrial and Urban Development Interests

Industrial Trade Associations
  National Association of Homebuilders
  Chambers of Commerce
  Tourism Boards

Environmental, Recreation, and Public Interest Organizations

Sierra Club
  Trout Unlimited
  American Rivers
  Native American Rights Fund
  Ecological Society of America
  National Wildlife Federation
  Nature Conservancy
  Natural Resources Defense Council
  Environmental Defense Fund

Court and Legal Systems

American Bar Association
  Environmental Law Institute

Scientific and Academic Organizations

American Association for the Advancement of Science
  International Water Resources Association
  National Academy of Sciences
  American Fisheries Society
  American Public Health Association
  American Water Resources Association
  American Society of Civil Engineers
  National Society of Professional Engineers
  American Society of Plumbing Engineers
  Universities Council for Water Resources

Suppliers and Consultants

American Council of Engineering Companies
  Water and Wastewater Equipment Manufacturers Association
  Trade associations for components like pumps, valves, etc.

Federal Agencies with Roles in Water Management

Federal Energy Regulatory Commission
  FEMA
  Department of Energy
  US Army Corps of Engineers
  Bureau of Reclamation
  Environmental Protection Agency
  Department of Health and Human Services
  Natural Resources Conservation Service
  US Forest Service
  Bureau of Land Management
  USDA offices involved in research and extension
  FmHA
  Department of Commerce
  HUD USGS
  National Weather Service (NOAA)
  National Park Service
  US Fish and Wildlife Service

Other Stakeholder Groups

Elected Officials and Appointed Boards

Boards of Water Management Agencies
   State Governors
   City Governments
   County Governments
   Legislators
   Councils of Government

State Government

Divisions of Water Resources/State Engineers
   Environmental and Health Agencies
   Natural Resources Departments
   Public Utility Commissions
   Agriculture Departments
   Forest Services
   Mining Boards and Agencies
   Water and Geological Surveys
   Climatologists
   Coastal Management Agencies
   Fish and Wildlife Departments
   Parks and Recreation Departments

Local and Regional Water Agencies

River Basin Authorities and Water Districts
   Municipal Water Agencies
   Municipal Wastewater Agencies
   Local Stormwater Agencies
   Water and Sanitation Districts
   Soil and Water Districts
   Flood Districts

Hydropower, Navigation, Flood Control Interests

TVA
   Navigation Companies
   Electric Power Utilities
   Port Authorities

Agriculture and Resource Development Sectors

Farmers and Irrigators
  Livestock Interests
  Miners
  Forest/Loggers
  Government Agriculture Departments
  State Boards of Agriculture
  Ditch Companies

Industrial and Urban Development Interests

State Economic Development Agencies
  Manufacturing Industries
  Realtors

Environmental, Recreation, and Public Interest Organizations

Environmental Groups
  Recreationists
  Wildlife Enthusiasts
  Wilderness Advocates
  Species Preservation Groups

Court and Legal Systems

Federal Court System, including Supreme Court
  Department of Justice
  State Courts including Water Courts
  State Attorney Generals
  Attorneys

Universities, Research Institutes, Publishers

Water-Related Departments of Colleges and Universities
  State Water Resources Research Institutes
  Government Research Offices
  Policy Research Centers
  Publishers
  University Experiment Stations and Extension Offices
  Groups Involved in Environmental and Water Education

Contractors

Utility Contractors
   Heavy Contractors
   Water Well Contractors

Suppliers

Manufacturers and suppliers of equipment and services used in water industry
   (Examples include: pumping, valves, piping, instruments, chemicals, treatment equipment, IT–based devices such as utility locators, software, and others)

Consultants

Engineering and other consultants

Financiers

Banks
   Bonding Agencies
   State Financing Authorities

Appendix D

Water Journals

Journal information is available by Internet searches by the titles

Advances in Water Resources
African Journal of Aquatic Science
Agricultural Water Management
Applied Water Science
Aqua
Aquatic Geochemistry
Aquatic Sciences
Australasian Journal of Water Resources
AWWA Water Science
Blue-Green Systems
Canadian Water Resources Journal
Climate Risk Management
Desalination
Desalination and Water Treatment
Discover Water
Ecohydrology
Ecohydrology & Hydrobiology
EOS, Trans AGU
Estuaries and Coasts
Flow Measurement and Instrumentation
Frontiers in Water
Groundwater
Groundwater Monitoring and Remediation
H_2O Open
Hydrogeology

Water Resources Management: Principles, Methods, and Tools, First Edition. Neil Grigg.
© 2023 John Wiley & Sons, Inc. Published 2023 by John Wiley & Sons, Inc.

Hydroinformatics
Hydrological Processes
Hydrological Sciences Journal
Hydrology
Hydrology Research
IDA Journal of desalination and water reuse
Ingeniería del Agua
Inland Waters
International Aquatic Research
International Journal of Energy and Water Resources
International Journal of River Basin Management
International Journal of WR Development
International Review of Hydrobiology
Irrigation and Drainage
Irrigation Science
ISH Journal of Hydraulic Engineering
Journal of Applied Water Engineering In Research
Journal of Contaminant Hydrology
Journal of Contemporary Water Research & Education
Journal of Environmental Engineering ASCE
Journal of Flood Risk Management
Journal of Hydraulic engineering
Journal of Hydraulic Research
Journal of Hydro-environment Research
Journal of Hydroinformatics
Journal of Hydrologic engineering
Journal of Hydrology
Journal of Hydrology: Regional Studies
Journal of Irrigation And Drainage Engineering
Journal of Pipeline Systems Engineering And Practice
Journal of Sustainable Water In The Built Environment
Journal of the American Water Resources Association
Journal of Water and Health
Journal of Water Chemistry And Technology
Journal of Water Process Engineering
Journal of Water Resources Planning and Management
Journal of Water Reuse and Desalination
Journal of Water Sanitation and Hygiene for Development
Journal of Water Supply: Research and Technology-AQUA
Journal of Water, Sanitation and Hygiene for Development
Journal, American Water Works Association

Lake and Reservoir Management
Lakes & Reservoirs: Science, Policy and Management for Sustainable Use
Limnology and Oceanography
Mine Water and the Environment
Paddy and Water Environment
Ribaqua Revista Iberoamericana del Agua
River Research and Applications
Sustainability of Water Quality and Ecology
Sustainable Water Resources Management
Urban Water Journal
Water
Water and Climate Change
Water and Environment Journal
Water and Health
Water Conservation Science and Engineering
Water Economics and Policy
Water Environment Research
Water History
Water International
Water Policy
Water Practice and Technology
Water Quality Research Journal
Water Quality Research Journal of Canada
Water Research
Water resources
Water Resources & Industry
Water Resources and Economics
Water Resources and Industry
Water Resources and Rural Development
Water Resources Management
Water Resources Research
Water Reuse
Water Science
Water Science & Technology
Water Security
Water Supply
Water Waves
Water, Air, and Soil Pollution
Wetlands
WIREs Water
World Water Policy

Appendix E

Glossary of Water Management Terms

		Source[a]
Acre-foot	Volume of water required to cover 1 acre of land ($43\,560\,\text{feet}^2$) to a depth of 1 foot. Equal to 325 851 gal or $1233\,\text{m}^3$.	3
Algal blooms	Sudden spurts of algal growth, which can degrade water quality and indicate potentially hazardous changes in local water chemistry.	2
Alluvium	Deposits of clay, silt, sand, gravel, or other particulate material that has been deposited by a stream or other body of running water in a streambed, on a flood plain, on a delta, or at the base of a mountain.	3
Amortization	Process of repaying debt by making periodic payments that include both principal and interest	
Appropriation	The setting aside of money by Congress, through legislation, for a specific use.	4
Appropriation doctrine	System for allocating water to private individuals used in most Western states. Contrasts with Riparian Water Rights.	3
Aquifer	A natural underground layer, often of sand or gravel, that contains water.	1
Aquifer (confined)	Soil or rock below the land surface that is saturated with water. There are layers of impermeable material both above and below it and it is under pressure so that when the aquifer is penetrated by a well, the water will rise above the top of the aquifer.	3
Aquifer (unconfined)	An aquifer whose upper water surface (water table) is at atmospheric pressure, and thus is able to rise and fall.	3

(Continued)

Water Resources Management: Principles, Methods, and Tools, First Edition. Neil Grigg.
© 2023 John Wiley & Sons, Inc. Published 2023 by John Wiley & Sons, Inc.

		Source[a]
Arbitrage	Process of investing funds borrowed from one source to yield higher interest returns.	
Artesian water	Ground water that is under pressure when tapped by a well and is able to rise above the level at which it is first encountered. See aquifer (confined).	3
Artificial recharge	Process where water is put back into ground-water storage from surface-water supplies such as irrigation, or induced infiltration from streams or wells.	3
Authorization	House and Senate Public Works Committee resolutions or specific legislation, which provides the legal basis for projects. The money necessary for accomplishing the work is not a part of the authorization, but must come from an appropriation by Congress.	4
Basin	(i) Drainage area of a lake or stream as river basin. (ii) A naturally or artificially enclosed harbor for small craft as yacht basin.	4
Benthic organisms	The worms, clams, crustaceans, and other organisms that live at the bottom of the estuaries and the sea.	2
Best available technology	The water treatment(s) that EPA certifies to be the most effective for removing a contaminant.	1
Biochemical oxygen demand	The amount of dissolved oxygen in parts per million required by organisms to enable them to decompose the organic matter present in the water.	4
Biodiversity	The variety and variability among living organisms and the ecological complexes in which they occur. Diversity can be defined as the number of different items and their relative frequencies. The term encompasses three basic levels of biodiversity: ecosystems, species, and genes.	2
Biomonitoring	The assessment of human exposure to chemicals by measuring the chemicals or their metabolites (breakdown products) in human tissues or fluids such as blood or urine. Blood and urine levels reflect the amount of the chemical in the environment that actually gets into the body.	2
Bond	Evidence of debt, with fixed interest rates, maturity dates, and amount.	
Cathodic protection	A method that uses electrical current for corrosion control.	
Chronic health effect	The possible result of exposure over many years to a drinking water contaminant at levels above its MCL.	1
Coliform	A group of related bacteria whose presence in drinking water may indicate contamination by disease-causing microorganisms.	1

		Source[a]
Commercial water use	Water used for motels, hotels, restaurants, office buildings, other commercial facilities, and institutions. Water for commercial uses comes both from public-supplied sources, such as a county water department, and self-supplied sources, such as local wells.	3
Community water system	A water system which supplies drinking water to 25 or more of the same people year-round in their residences.	1
Compliance	The act of meeting all state and federal drinking water regulations.	1
Confluence	The place where streams meet.	4
Consumptive use	That part of water withdrawn that is evaporated, transpired by plants, incorporated into products or crops, consumed by humans or livestock, or otherwise removed from the immediate water environment. Also referred to as water consumed.	3
Contaminant	Anything found in water (including microorganisms, minerals, chemicals, radionuclides, etc.) that may be harmful to human health.	1
Conveyance loss	Water lost in transit from a pipe, canal, or ditch by leakage or evaporation. Generally, the water is not available for further use; however, leakage from an irrigation ditch, for example, may percolate to a ground-water source and be available for further use.	3
Coordination	The process of bringing harmony to diverse interests, as in integrating the view of water stakeholders in a decision process.	
Corrosion control	Control of galvanic corrosion or passivation of metal release in pipes	
Cryptosporidium	A microorganism commonly found in lakes and rivers which is highly resistant to disinfection. Cryptosporidium has caused several large outbreaks of gastrointestinal illness, with symptoms that include diarrhea, nausea, and/or stomach cramps.	1
Cubic feet per second (cfs)	A rate of flow. One "cfs" is equal to 7.48 gal of water flowing each second.	
Dam	A barrier constructed across a valley for impounding water or creating a reservoir.	4
Debt limits	Limits on amount of debt imposed by a legal instrument such as a constitution.	
Debt service	Principal and interest charges on a debt.	

(Continued)

		Source[a]
Degree of protection	The amount of protection that a flood control measure is designed for, as determined by engineering feasibility, economic criteria, social, environmental, and other considerations.	4
Desalinization	Removal of salts from saline water to provide freshwater. This method is becoming a more popular way of providing freshwater to populations.	3
Dike	An embankment to confine or control water, and/or soil.	4
Discharge	Volume of water that passes a given location within a given period of time. Usually expressed in cubic feet per second.	3
Disinfection by-product	A compound formed by the reaction of a disinfectant such as chlorine with organic material in the water supply; a chemical byproduct of the disinfection process.	2
Distribution system	A network of pipes leading from a treatment plant to customers' plumbing systems.	1
Diversion channel	(i) An artificial channel constructed around a town or other point of high potential flood damages to divert floodwater from the main channel to minimize flood damages. (ii) A channel carrying water from a diversion dam.	4
Domestic water use	Water used for household purposes, such as drinking, food preparation, bathing, washing clothes, dishes, and dogs, flushing toilets, and watering lawns and gardens.	3
Drainage basin	Land area (from) where precipitation runs off into streams, rivers, lakes, and reservoirs. It is a land feature that can be identified by tracing a line along the highest elevations between two areas on a map, often a ridge. Also called a "watershed."	3
Drip irrigation	An irrigation method where pipes or tubes filled with water slowly drip onto crops. Drip irrigation is a low-pressure method of irrigation and less water is lost to evaporation than other methods.	3
Ecological indicators	Measurable characteristics related to the structure, composition, or functioning of ecological systems; a measure, an index of measures, or a model that characterizes an ecosystem or one of its critical components.	2
Ecological processes	The metabolic functions of ecosystems – energy flow, elemental cycling, and the production, consumption, and decomposition of organic matter.	2

		Source[a]
Ecosystem	The interacting system of a biological community and its nonliving environmental surroundings. A geographic area including all living organisms (people, plants, animals, and microorganisms), their physical surroundings (such as soil, water, and air), and the natural cycles that sustain them.	2
Effluent	Water that flows from a sewage treatment plant after it has been treated.	3
Environmental assessment	A planning report which presents the first thorough examination of alternative plans that positively demonstrate that the environmental and social consequences of a federal action were considered.	4
Environmental impact statement	A report required by section 102(2)(c) of Public Law 91-190 for all federal actions which significantly impact on the quality of the human environment or are environmentally controversial.	4
Environmental indicators	Scientific measurements that help measure over time the state of air, water, and land resources, pressures on those resources, and resulting effects on ecological condition and human health. Indicators show progress in making the air cleaner and the water purer and in protecting land.	2
Erosion	The wearing away of land surface by wind or water, intensified by land-clearing practices related to farming, residential or industrial development, road building, or logging.	2
Estuary	A place where fresh and salt water mix, such as a bay, salt marsh, or where a river enters an ocean.	3
Eutrophication	The slow aging process during which a lake, estuary, or bay evolves into a bog or marsh and eventually disappears. During the later stages of this process, the water body is choked by abundant plant life that result from higher levels of nutritive compounds such as nitrogen and phosphorus.	2
Evaporation	The process of liquid water becoming water vapor, including vaporization from water surfaces, land surfaces, and snow fields, but not from leaf surfaces.	3
Evapotranspiration	The sum of evaporation and transpiration.	3
Finished water	Water that has been treated and is ready to be delivered to customers.	1

(Continued)

		Source[a]
Fish kill	A large-scale die-off of fish caused by factors such as pollution, noxious algae, harmful bacteria, and hypoxic conditions.	2
Flood	An overflow of water onto lands that are used or usable by man and not normally covered by water.	3
Flood (1% flood)	This is the same as a 100-year flood and is a flood which has a 1% chance of occurrence in any year.	4
Flood plain	A strip of relatively flat and normally dry land alongside a stream, river, or lake that is covered by water during a flood.	3
Flood, 100-year	A 100-year flood does not refer to a flood that occurs once every 100 years, but to a flood level with a 1% chance of being equaled or exceeded in any given year.	3
Floodproofing	Techniques for preventing flood damage to the structure and contents of buildings in a flood hazard area.	4
Floodwall	A wall, usually built of reinforced concrete, to confine streamflow to prevent flooding.	4
Freshwater	Water that contains less than 1000 milligrams per liter (mg/l) of dissolved solids; generally, more than 500 mg/l of dissolved solids is undesirable for drinking and many industrial uses.	3
Gage height	Height of the water surface above the gage datum (zero point). Gage height is often used interchangeably with the more general term, stage.	3
Gaging station	A site on a stream, lake, reservoir or other body of water where observations and hydrologic data are obtained.	3
General obligation bond	Bond which is secured by taxing power.	
Giardia lamblia	A microorganism frequently found in rivers and lakes, which, if not treated properly, may cause diarrhea, fatigue, and cramps after ingestion.	1
Greywater	Wastewater from clothes washing machines, showers, bathtubs, hand washing, lavatories, and sinks.	3
Ground water	Subsurface water that occurs beneath the water table in soils and geologic formations that are fully saturated.	2
Ground water recharge	Inflow of water to a ground-water reservoir from the surface. Infiltration of precipitation and its movement to the water table is one form of natural recharge.	3
Ground water, confined	See aquifer (confined).	3
Ground water, unconfined	See aquifer (unconfined).	3

		Source[a]
Habitat	The place where a population (e.g., human, animal, plant, and microorganism) lives and its surroundings, both living and nonliving.	2
Hardness	A water-quality indication of the concentration of alkaline salts in water, mainly calcium and magnesium. If the water you use is "hard" then more soap, detergent or shampoo is necessary to raise a lather.	3
Hazardous waste	Byproducts of society that can pose a substantial or potential threat to human health or the environment when improperly managed. Hazardous waste possesses at least one of four characteristics: ignitability, corrosivity, reactivity, or toxicity.	2
Headwaters	(i) The upper reaches of a stream near its source. (ii) The region where groundwaters emerge to form a surface stream. (iii) The water upstream from a structure.	4
Heavy metals	Metallic elements with high atomic weights (e.g., mercury, chromium, cadmium, arsenic, and lead); can damage living things at low concentrations and tend to accumulate in the food chain.	2
Herbicide	A form of pesticide used to control weeds that limit or inhibit the growth of the desired crop.	2
Hydrologic cycle	Movement or exchange of water between the atmosphere and earth.	2
Hypoxia/hypoxic waters	Waters with low levels of dissolved oxygen concentrations, typically less than 2 ppm, the level generally accepted as the minimum required for most marine life to survive and reproduce.	2
Industrial water use	Water used for industrial purposes in such industries as steel, chemical, paper, and petroleum refining.	3
Infiltration	Flow of water from the land surface into the subsurface.	3
Injection well	A well constructed for the purpose of injecting treated wastewater directly into the ground.	3
Inorganic contaminants	Mineral-based compounds such as metals, nitrates, and asbestos.	1
Integrated water management	Water management that responds in a balanced way to the needs and viewpoints of stakeholders.	
Intercepting sewer	A conduit that receives flow from a number of transverse sewers or outlets and conducts such waters to a point for treatment or disposal.	4
Irrigation water use	Water application on lands to assist in the growing of crops and pastures or to maintain vegetative growth in recreational lands, such as parks and golf courses.	3

(Continued)

		Source[a]
Leaching	Process by which soluble materials in the soil, such as salts, nutrients, pesticide chemicals or contaminants, are washed into a lower layer of soil or are dissolved and carried away by water.	3
Levee	A natural or manmade earthen barrier along the edge of a stream, lake, or river. Land alongside rivers can be protected from flooding by levees.	3
Livestock water use	Water used for livestock watering, feed lots, dairy operations, fish farming, and other on-farm needs.	3
Lock	An enclosed part of a canal, waterway, etc., equipped with gates so that the level of the water can be changed to raise or lower from one level to another.	4
Man-made water resources system	Set of constructed facilities that control water flow and quality.	
Maximum contaminant level (MCL)	The highest level of a contaminant that EPA allows in drinking water. MCLs ensure that drinking water does not pose either a short- or long-term health risk. EPA sets MCLs at levels that are economically and technologically feasible.	1
Microorganisms	Tiny living organisms that can be seen only with the aid of a microscope. Also known as microbes.	1
Milligrams per liter (mg/l)	An unit of the concentration of a constituent in water or wastewater. It represents 0.001 g of a constituent in 1 l of water. It is approximately equal to one part per million (PPM).	3
Million gallons per day (mgd)	Rate of flow of water equal to $1.5472 \, feet^3/s$, or $3.0689 \, acre\text{-}feet/day$.	3
Mining water use	Water use during quarrying rocks and extracting minerals from the land.	3
Monitoring	Testing that water systems must perform to detect and measure contaminants.	1
Municipal bond	Bond issued by a municipality, sometimes called a tax exempt bond.	
Municipal water system	A water system that has at least five service connections or which regularly serves 25 individuals for 60 days; also called a public water system.	3
Natural water resources system	Natural hydrologic elements that include the atmosphere, watersheds, stream channels, wetlands, floodplains, aquifers and groundwater systems, lakes, estuaries, seas, and the ocean.	
Nephelometric turbidity unit (NTU)	Unit of measure for the turbidity of water. Essentially, a measure of the cloudiness of water as measured by a nephelometer. Turbidity is based on the amount of light that is reflected off particles in the water.	3

		Source[a]
Nitrate	The primary chemical form of nitrogen in most aquatic systems; occurs naturally; a plant nutrient and fertilizer; can be harmful to humans and animals.	2
Noncommunity water system	A public water system that is not a community water system. Nontransient noncommunity water systems are those that regularly supply water to at least 25 of the same people at least 6 months/year but not year-round (e.g., schools, factories, office buildings, and hospitals that have their own water systems). Transient noncommunity water systems provide water in a place where people do not remain for long periods of time (e.g., a gas station or campground).	2
Nonnative species	A species that has been introduced by human action, either intentionally or by accident, into areas outside its natural geographical range. Other names for these species include alien, exotic, introduced, and nonindigenous.	2
Non-point source (NPS) pollution	Pollution discharged over a wide land area, not from one specific location. Caused by sediment, nutrients, organic, and toxic substances originating from land-use activities, which are carried to lakes and streams by surface runoff.	3
Non-structural measures	Programs and activities for water management that do not require constructed facilities, such as pricing, zoning, education, regulatory programs, or insurance.	
Non-transient, Non-community Water system	A water system which supplies water to 25 or more of the same people at least 6 months/year in places other than their residences.	1
Nutrient	Any substance assimilated by living things that promotes growth. The term is generally applied to nitrogen and phosphorus, but is also applied to other essential and trace elements.	2
Organic contaminants	Carbon-based chemicals, such as solvents and pesticides, which can get into water through runoff from cropland or discharge from factories. EPA has set legal limits on 56 organic contaminants.	1
Organic matter	Plant and animal residues, or substances made by living organisms. All are based upon carbon compounds.	3
Osmosis	Movement of water molecules through a thin membrane. The osmosis process occurs in our bodies and is also one method of desalinizing saline water.	3
Outfall	Place where a sewer, drain, or stream discharges; the outlet or structure through which reclaimed water or treated effluent is finally discharged to a receiving water body.	3

(Continued)

		Source[a]
Oxygen demand	The need for molecular oxygen to meet the needs of biological and chemical processes in water.	3
Parts per billion	Number of "parts" by weight of a substance per billion parts of water. Used to measure extremely small concentrations.	3
Parts per million	Number of "parts" by weight of a substance per million parts of water. This unit is commonly used to represent pollutant concentrations.	3
Pathogen	A disease-causing organism.	1
Peak flow	Maximum instantaneous discharge of a stream or river at a given location. It usually occurs at or near the time of maximum stage.	3
Per capita use	Average amount of water used per person during a standard time period, generally per day.	3
Percolation	The movement of water through the openings in rock or soil, or entrance of a portion of the streamflow into the channel materials to contribute to ground water replenishment.	3
Permeability	Ability of a material to allow the passage of a liquid, such as water through rocks.	3
Pesticide	Any substance or mixture of substances intended to prevent, destroy, repel, or mitigate any pest. Pests can be insects, mice and other animals, unwanted plants (weeds), fungi, or microorganisms such as bacteria and viruses. Though often misunderstood to refer only to insecticides, the term "pesticide" also applies to herbicides, fungicides, and various other substances used to control pests. Under U.S. law, a pesticide is also any substance or mixture of substances intended for use as a plant regulator, defoliant, or desiccant.	2
pH	A measure of the relative acidity or alkalinity of water. Water with a pH of seven is neutral; lower pH levels indicate increasing acidity, while pH levels higher than seven indicate increasingly basic solutions.	3
Phosphorus	An essential chemical food element that can contribute to the eutrophication of lakes and other water bodies.	2
Photosynthesis	The manufacture by plants of carbohydrates and oxygen from carbon dioxide mediated by chlorophyll in the presence of sunlight.	2
Phytoplankton	That portion of the plankton community composed of tiny plants (e.g., algae, diatoms).	2

		Source[a]
Planners	Persons who prepare plans that lead to water management decisions. They often work in offices with responsibility for coordination, rather than direct implementation of projects.	
Point-source pollution	Water pollution coming from a single point, such as a sewage-outflow pipe.	3
Polychlorinated biphenyls (PCBs)	A group of synthetic, toxic industrial chemical compounds once used in making paint and electrical transformers, which are chemically inert and not biodegradable. PCBs were frequently found in industrial wastes, and subsequently found their way into surface and ground waters. As a result of their persistence, they tend to accumulate in the environment. In terms of streams and rivers, PCBs are drawn to sediment, to which they attach and can remain virtually indefinitely. Although virtually banned in 1979 with the passage of the Toxic Substances Control Act, they continue to appear in the flesh of fish and other animals.	3
Polywrap	A polyethylene encasement for ductile iron pipe from external corrosion.	
Porosity	A measure of the water-bearing capacity of subsurface rock. With respect to water movement, it is not just the total magnitude of porosity that is important, but the size of the voids and the extent to which they are interconnected, as the pores in a formation may be open, or interconnected, or closed and isolated. For example, clay may have a very high porosity with respect to potential water content, but it constitutes a poor medium as an aquifer because the pores are usually so small.	3
Potable water	Water of a quality suitable for drinking.	3
Precipitation	Rain, snow, hail, sleet, dew, and frost.	3
Primacy state	A State that has the responsibility and authority to administer EPA's drinking water regulations within its borders.	1
Primary wastewater treatment	The first stage of the wastewater-treatment process where mechanical methods, such as filters and scrapers, are used to remove pollutants. Solid material in sewage also settles out in this process.	1
Prior appropriation doctrine	System for allocating water to private individuals used in most Western states. See "appropriation doctrine."	3
Public supply	Water withdrawn by public governments and agencies, such as a city water department, and by private companies that is then delivered to users.	3

(Continued)

		Source[a]
Public water system (PWS)	Any water system which provides water to at least 25 people for at least 60 days annually.	1
Public water use	Water supplied from a public-water supply and used for such purposes as firefighting, street washing, and municipal parks and swimming pools.	3
Publicly owned treatment works	Water or wastewater treatment plants owned by public entities.	
Radionuclides	Any man-made or natural element that emits radiation and that may cause cancer after many years of exposure through drinking water.	1
Rating	Indication of the investment quality of the bond.	
Rating curve	A drawn curve showing the relation between gage height and discharge of a stream at a given gaging station.	3
Raw water	Water in its natural state, prior to any treatment for drinking.	1
RCRA hazardous waste	Applies to certain types of hazardous wastes that appear on EPA's regulatory listing (RCRA) or that exhibit specific characteristics of ignitability, corrosiveness, reactivity, or excessive toxicity.	2
Reach	A length, distance or leg of a channel or other watercourse.	4
Recharge	Water added to an aquifer. For instance, rainfall that seeps into the ground.	3
Reclaimed wastewater	Treated wastewater that can be used for beneficial purposes, such as irrigating certain plants.	3
Recurrence interval	The average time interval between actual occurrence of a flood of a given magnitude.	4
Recycled water	Water that is used more than one time before it passes back into the natural hydrologic system.	3
Regulators	Persons with responsibility to monitor and enforce compliance with laws and regulations.	
Rehabilitation	A major repair job. Usually involves considerable reconstruction of already existing structures.	4
Remediation	Cleanup or other methods used to remove or contain a toxic spill or hazardous materials from a contaminated site.	2
Renewal	Infrastructure repair, rehabilitation, or replacement.	
Replacement	Substitution of infrastructure component with a new one of equal or superior value and capacity.	
Reservoir	A pond, lake, or basin, either natural or artificial, for the storage, regulation, and control of water.	3

		Source[a]
Return flow	That part of a diverted flow that is not consumptively used and returned to its original source or another body of water. In case of irrigation, drainage water from irrigated farmlands.	3
Revenue bond	Bonds guaranteed by revenues of an operation rather than tax base.	
Reverse osmosis	Process of removing salts from water using a membrane. An advanced method of water or wastewater treatment that relies on a semi-permeable membrane to separate waters from pollutants.	3
Revetment	A facing of stone, concrete, sandbags, etc., to protect a bank of earth from erosion. A retaining wall.	4
Riparian water rights	Rights of an owner whose land abuts water.	3
Riprap	A layer, facing, or protective mound of randomly placed stones to prevent erosion, scour, or sloughing of a structure or embankment. The stone used for this purpose is also called riprap.	4
Risk	The probability that a health problem, injury, or disease will occur.	2
River	A natural stream of water of considerable volume, larger than a brook or creek.	3
River basin	A water resource basin is a portion of a water resource region defined by a hydrological boundary which is usually the drainage area of one of the lesser streams in the region.	4
Runoff	That part of precipitation, snowmelt, or irrigation water that runs off the land into streams or other surface water. It can carry pollutants from the air and land into receiving waters.	2
Saline water	Water that contains significant amounts of dissolved solids: Fresh water – less than 1000 parts per million (ppm); Slightly saline water – from 1000 to 3000 ppm; Moderately saline water – from 3000 to 10 000 ppm; Highly saline water – from 10 000 to 35 000 ppm.	3
Sample	Water analyzed for the presence of EPA-regulated drinking water contaminants.	1
Secondary drinking water standards	Non-enforceable federal guidelines regarding cosmetic effects (such as tooth or skin discoloration) or aesthetic effects (such as taste, odor, or color) of drinking water.	1

(Continued)

		Source[a]
Secondary wastewater treatment	Treatment (following primary wastewater treatment) involving the biological process of reducing suspended, colloidal, and dissolved organic matter in effluent from primary treatment systems and which generally removes 80–95% of the Biochemical Oxygen Demand (BOD) and suspended matter.	3
Sediment	Material in suspension in water or recently deposited from suspension. In the plural the word is applied to all kinds of deposits from the waters of streams, lakes, or seas.	3
Seepage	(i) The slow movement of water through small cracks, pores, Interstices, etc., of a material into or out of a body of surface or subsurface water. (ii) The loss of water by infiltration into the soil from a canal, ditches, laterals, watercourse, reservoir, storage facilities, or other body of water, or from a field.	3
Self-supplied water	Water withdrawn from a surface- or ground-water source by a user rather than being obtained from a public supply. An example would be homeowners getting their water from their own well.	3
Septic tank	Tank used to detain domestic wastes to allow the settling of solids prior to distribution to a leach field for soil absorption.	3
Service providers	Utilities, agencies, or other organizations that provide the water service(s) required by customers.	
Sewage treatment plant	Facility designed to receive the wastewater from domestic sources and to remove materials that damage water quality and threaten public health and safety when discharged into receiving streams or bodies of water.	3
Sewer	System of underground pipes that collect and deliver wastewater to treatment facilities or streams.	3
Sludge	Solid, semisolid, or liquid waste generated from a municipal, commercial, or industrial wastewater facility.	2
Sole source aquifer	An aquifer that supplies 50% or more of the drinking water of an area.	1
Source water	Water in its natural state, prior to any treatment for drinking.	1
Spillway	A waterway or a dam or other hydraulic structure used to discharge excess water to avoid overtopping of a dam.	4
Stakeholder	Person or group having an interest in outcome of a water-related decision.	

		Source[a]
Standard project flood	A flood that may be expected from the most severe combination of meteorological and hydrological conditions that are reasonably characteristic of the geographical region involved, excluding extremely rare combinations.	4
Storm sewer	Sewer that carries only surface runoff, street wash, and snow melt from the land. In a separate sewer system, storm sewers are completely separate from those that carry domestic and commercial wastewater (sanitary sewers).	3
Stream	General term for a body of flowing water.	3
Streamflow	Water discharge that occurs in a natural channel.	3
Structural measures	Constructed facilities used to control water flow and quality.	
Subsidence	Dropping of the land surface as a result of ground water being pumped.	3
Superfund	Program operated under the legislative authority of the Comprehensive Environmental Response, Compensation, and Liability Act (CERCLA) and the Superfund Amendments and Reauthorization Act (SARA) that funds and carries out EPA solid waste emergency and long-term removal and remedial activities.	2
Support sector	Sets of individuals and organizations that provide business or decision making support to the service providers, regulators, and coordinators of the water industry.	
Surface water	The water that systems pump and treat from sources open to the atmosphere, such as rivers, lakes, and reservoirs.	1
Suspended sediment	Very fine soil particles that remain in suspension in water for a considerable period of time without contact with the bottom. Such material remains in suspension due to the upward components of turbulence and currents and/or by suspension.	3
Suspended solids	Solids that are not in true solution and that can be removed by filtration. Such suspended solids usually contribute directly to turbidity. Defined in waste management, these are small particles of solid pollutants that resist separation by conventional methods.	3
Tax increment financing	Using increase in taxes to repay bonds or other debt incurred to build facilities.	

(Continued)

		Source[a]
Tertiary wastewater treatment	Selected biological, physical, and chemical separation processes to remove organic and inorganic substances that resist conventional treatment practices; the additional treatment of effluent beyond that of primary and secondary treatment methods to obtain a very high quality of effluent.	3
Thermal pollution	Reduction in water quality caused by increasing its temperature, often due to disposal of waste heat from industrial or power generation processes. Thermally polluted water can harm the environment because plants and animals can have a hard time adapting to it.	3
Thermoelectric power water use	Water used in the process of the generation of thermoelectric power. Power plants that burn coal and oil are examples of thermoelectric-power facilities.	3
Threatened and endangered species	Those species that are in danger of extinction throughout all or a significant portion of their range or are likely to become endangered in the future.	2
Transient, non-community water system	A water system which provides water in a place such as a gas station or campground where people do not remain for long periods of time.	1
Transmissibility (ground water)	Capacity of a rock to transmit water under pressure. The coefficient of transmissibility is the rate of flow of water, at the prevailing water temperature, in gallons per day, through a vertical strip of the aquifer one foot wide, extending the full saturated height of the aquifer under a hydraulic gradient of 100%. A hydraulic gradient of 100% means a one foot drop in head in one foot of flow distance.	3
Transpiration	Process by which water that is absorbed by plants, usually through the roots, is evaporated into the atmosphere from the plant surface, such as leaf pores. See evapotranspiration.	3
Tributary	A smaller river or stream that flows into a larger river or stream. Usually, a number of smaller tributaries merge to form a river.	3
Tuberculation	A phenomena whereby corrosion products and precipitated solids accumulate on the interior wall of a pipe.	
Turbidity	The cloudy appearance of water caused by the presence of tiny particles. High levels of turbidity may interfere with proper water treatment and monitoring.	1
Underground storage tanks	Tanks and their underground piping that have at least 10% of their combined volume underground.	2
Underwriter	Broker or bank that markets bonds for a fee buys bonds through a sale agreement.	

		Source[a]
Unsaturated zone	The zone immediately below the land surface where the pores contain both water and air, but are not totally saturated with water. These zones differ from an aquifer, where the pores are saturated with water.	3
Urban water services	Water supply, wastewater, and stormwater management services in urban areas.	
Variance	State or EPA permission not to meet a certain drinking water standard.	1
Vulnerability assessment	An evaluation of drinking water source quality and its vulnerability to contamination by pathogens and toxic chemicals.	1
Wastewater	Water that has been used in homes, industries, and businesses that is not for reuse unless it is treated.	3
Water industry	The utilities, agencies, firms, interest groups and organizations that deliver water services, regulate the water environment, and provide coordination and support functions.	
Water management	The planning, organization, and control of structural and nonstructural water systems for beneficial purposes.	
Water quality	A term used to describe the chemical, physical, and biological characteristics of water, usually in respect to its suitability for a particular purpose.	3
Water quality criteria	Levels of water quality expected to render a body of water suitable for its designated use. Criteria are based on specific levels of pollutants that would make the water harmful if used for drinking, swimming, irrigation, fish production, or industrial processes.	2
Water quality standards	State-adopted and EPA-approved ambient standards for water bodies. The standards define the water quality goals of a water body by designating the uses of the water and setting criteria to protect those uses. The standards protect public health and welfare, enhance the quality of the water, and provide the baseline for surface water protection under the Clean Water Act.	2
Water resources system	Combination of control facilities and environmental elements that work together to achieve water management purposes.	
Water resources systems management	Application of structural and nonstructural measures to control natural and man-made water resources systems for beneficial purposes.	
Water table	Top of the water surface in the saturated part of an aquifer.	3

(Continued)

		Source[a]
Water use	Water that is used for a specific purpose, such as for domestic use, irrigation, or industrial processing. Water use pertains to human's interaction with and influence on the hydrologic cycle, and includes elements, such as water withdrawal from surface- and ground-water sources, water delivery to homes and businesses, consumptive use of water, water released from wastewater-treatment plants, water returned to the environment, and instream uses, such as using water to produce hydroelectric power.	3
Waterborne disease outbreak	The significant occurrence of acute illness associated with drinking water from a public water system or exposure encountered in recreational or occupational settings as determined by appropriate local or state agencies. (The Centers for Disease Control and Prevention defines an outbreak as two or more cases associated with drinking water as the route of exposure.)	2
Watershed	The land area from which water drains into a stream, river, or reservoir.	1
Well (water)	An artificial excavation put down by any method for the purposes of withdrawing water from the underground aquifers. A bored, drilled, or driven shaft, or a dug hole whose depth is greater than the largest surface dimension and whose purpose is to reach underground water supplies or oil, or to store or bury fluids below ground.	3
Wellhead protection area	The area surrounding a drinking water well or well field which is protected to prevent contamination of the well(s).	1
Wetland ecosystems	Areas that are inundated or saturated by surface or ground water at a frequency and duration sufficient to support, and that under normal circumstances do support, a prevalence of vegetation typically adapted for life in saturated soil conditions. Wetlands generally include swamps, marshes, bogs, and similar areas.	2
Withdrawal	Water removed from a ground- or surface-water source for use.	3
Xeriscape	Method of landscaping that uses plants that are well adapted to the local area and are drought-resistant.	3

[a] Terms are from publications and sources noted. In some cases definitions have been edited or paraphrased. If no source is indicated, the definitions are original here.

1. EPA's Drinking Water Glossary; 2. EPA Environmental Indicators Initiative; 3. USGS. Water Science Glossary of Terms; 4. Interagency Floodplain Management Review Committee. 1994. Sharing the Challenge: Floodplain Management into the twenty-first century. Washington, June.

Index

a

acre-foot 377
Administrative Doctrine 296
Administrative Procedures Act (APA) 291
affordability 228
Agricultural Research Service (ARS) 139
agricultural sector 47
agriculture and resource development
        sectors 368, 371
Albemarle-Pamlico National Estuary
        Partnership 334
algal blooms 377
allocation, of groundwater 297
alluvium 377
American Water Works Association
        (AWWA) 19, 160
amortization 377
ancient settlements 10
appropriation 377
appropriation doctrine 295, 377
aquifer 77, 377
arbitrage 378
arsenic 78
artesian water 378
artificial recharge 378
Asian Development Bank 205
asset management 150, 155–159, 347
Association of State Dam Safety Officials
        (ASDSO) 317
authorization 378
average precipitation and evaporation 84

b

basin 195, 378
benefit-cost analysis (BCA) 198, 241, 257
benthic organisms 378
big dams 10
biochemical oxygen demand 378
biodiversity 237–238, 378
biological diversity 236–237
biomonitoring 378
bio-terrorism 329

c

California Department of Water
        Resources 180
canals and pipelines 37–38
Capacity, Management, Operation, and
        Maintenance program (CMOM) 33
capital budgeting 195
capital budgets 205
capital improvement programs 205
carbonates 78
careers in, water resources management
        352–354
    asset management studies 347
    civil and environmental engineers 349
    construction and maintenance managers 350
    estuary recovery 348
    executives and managers of water agencies
        and utilities 349
    failures and remedial actions 348
    floodplain planning and management 348

Water Resources Management: Principles, Methods, and Tools, First Edition. Neil Grigg.
© 2023 John Wiley & Sons, Inc. Published 2023 by John Wiley & Sons, Inc.

careers in, water resources management (*cont'd*)
  green infrastructure   347
  hydrologists and ecologists   349
  information technology specialists   351
  landscape, irrigation, and water
    conservation specialists   351
  management consultants   350
  managers of financial, administrative, and
    legal affairs   352
  master infrastructure development
    plans   346
  multipurpose reservoirs   348
  permitting   348
  physical and life scientists   350–351
  planners and policy specialists   352
  planning infrastructure projects   347
  public information and education
    specialists   351
  rate increases, for wastewater utilities   347
  stormwater utility management   347
  system and process engineers   351
  technical sales and marketing   352
  wastewater plant operation and
    recycling   347
  water industry employers   345–346
  water recycling systems   347
  watershed manager   349–350
  water system operators   350
cash flow diagrams   249
Chancay–Lambayeque basin   339–340
channel flow   85–86
chemicals   120
Chesapeake Bay   334–335
chronic health effect   378
chronology   190–191
civil and environmental engineers   349
civil conflicts   322
civil society groups   367
Clean Water Act (CWA)   11, 16, 19, 54, 123,
    161, 216, 292, 294, 298, 306–307, 316
Clean Water State Revolving Fund   217, 298
climate change   69
  non-stationarity   69
  and population growth   4
climate shifts   322–323
coastal works   37

Code of Federal Regulations (CFR)   291
coliform   378
Colorado instream flow law   299
Colorado River Compact   18, 300
Colorado Water Conservation Board   299
commercial water use   379
community water systems   181, 379
compliance   379
comprehensive financial analysis   285
computer-based sustainability assessment
    frameworks   247
Congressional Research Service   176
conservation rate   63
construction and maintenance managers   350
consultants   372
consumptive use   379
contaminant   379
contractors   372
conventional fossil sources   13
conversion factors   357–358
conveyance loss   379
corrosion   160
corrosion control   379
cost allocation   288
cost data   287–288
cost effectiveness analysis   241
Council on Environmental Quality (CEQ)   238
court and legal systems   368, 371
crop failure   4
cryptosporidium   379
customer classes   51
customer expectations   197
customer rate classes   287
custom weather models   140
cyberattacks   320, 329
cybersecurity   39, 154

d

dam breaks   114
dam failures   309, 321
dams   13–14
dam safety   317–318
Darcy–Weisbach equation   20
data analytics and modeling   133
debt limits   379
debt service   379

decision support system   135
degree of protection   380
demand
  crop water demand   63–64
  peak demands   63
  for water and water infrastructure
    categories of water use   50–52
    direct requirements   46
    environmental water and instream
      flows   56–57
    flood control   56
    industrial and thermoelectric
      water uses   54
    irrigation   52–54
    sectors   46–47
    stormwater management   55–56
    wastewater management   54–55
    water use data   47–48
  for water treatment   62
Denver Union Water Company   19
depreciation and deferred maintenance   259
desalinization   380
detention pond   117
dike   380
disaster preparedness   326
disinfection by-product   380
distribution systems   31, 380
diversion channel   380
documentation   151
domestic and commercial water supplies   51
domestic water use   380
DPSIR framework   120
drainage
  basin   380
  and flood control law   308
  and flood risk management   292
drip irrigation   380
drip/trickle   36
droughts   322–323

e
ecohydrology   68
ecological processes   380
economic assessment   196
ecosystem   381
elected officials and appointed boards   367, 370

Electric Consumers Protection Act   13
emergency management   326
emergency preparedness   326
Endangered Species Act (ESA)   292, 294, 299
energy and communications   9
energy generation   43, 258
environmental and health issues   36
environmental assessment   198, 381
environmental degradation   30, 120
environmental feasibility   207
environmental impact statement   381
environmental indicators   236–237, 381
environmental integrity   119, 120, 126–129
environmental justice   224
environmental laws   289, 299
Environmental Protection Agency
    (USEPA)   179
environmental, recreation, and public interest
    organizations   368, 371
environmental water   292
environmental water quality   229
Envision   239
equality   224
equity   224
Erie Canal   13
estuary recovery   348
ethics   224
Europe   10
European Union   176
eutrophication   381
evaporation   85
evapotranspiration   381
executives and managers, of water agencies
    and utilities   349
external customers   287

f
failures and remedial actions   348
F-diagram   124
feasibility   199, 207
federal agencies   292, 369
Federal Emergency Management Agency
    (FEMA)   179
Federal Power Act (FPA)   292, 293
Federal Reclamation Act   17
Federal Rules of Evidence   301

Federal Water Power Act   12
financiers   372
financing and funding
  accounting and analysis   269–270
  affordability, of water services   279–280
  connection charges for system
      expansion   267
  debt financing   267–268
  financial assistance programs   268
  flood risk management   276–277
  instream flow management   277
  irrigation systems   276
  joint projects   270
  model of utility finance   265–266
  plans and budgets   264
  private water companies   277–278
  rate-setting   266–267
  stormwater   274–276
  wastewater management   273–274
  for water management   262–264
  water management agencies   277
  water supply   270–273
finished water   381
Fish and Wildlife Service environmental
      data   137
fish kill   382
fixture units   55
flood control   16, 182
Flood Control Act   17, 251, 313
flood control interests   368
flood control purposes   25
flood control services   46
flood insurance   310
Flood Insurance Act   11, 292, 313
Flood Insurance and Mitigation
      Administration   113
flood litigation   302
flood plain   382
floodplain planning and management   348
floodproofing   382
flood reservoir   117
flood risk statistics   310
floods
  damages   113–114
  dam breaks   114
  environmental benefits of floods   114–115
  flood frequency analysis   104–105

flood-related law   312–315
  forecasting   112–113
  hydraulics and conveyance   108–109
  hydrologic causes of   103–104
  mapping   113
  reservoir storage-routing   109–112
  risk   118
  storm and flood runoff   105–106
  threat causes and effects   311–312
  and water management   103
floodwall   382
flow and precipitation data   134
food shortages   4
forecasting   112–113
freshwater   382

g

gage height   382
gaging station   88, 382
Gardner, Charles   317
general obligation bond   382
geographical information systems (GIS)   139
Giardia lamblia   382
Global Water Partnership (GWP)   332
government agencies   10
government relief   4
Great Depression   191, 341
Great Mississippi River Floods   313
green infrastructure   347
Green Mountain Reservoir   341–342
greywater   382
groundwater   14
  confined   382
  law   297
  recharge   382
  unconfined   382

h

Harington, John   15
Harvard Water Program   134
hazardous waste   383
HAZUS Technical Manual   113
headwaters   383
health and safety   292
heavy metals   383
herbicide   383

Hoover, Herbert   191
horsepower of pump   42
human rights   242
Hurricane Katrina   228, 313
hydraulics and conveyance   108–109
hydroelectric energy   43
hydroelectric generation   25
hydroelectricity   257
hydrograph   106–107, 118
hydrologic cycle   383
hydrology and water management
  aquifers and ground water systems   76–79
  ecohydrology   68, 79–80
  hydrometeorology and climate   69–71
  low flow hydrology   73–75
  low flows and drought conditions   67
  physical hydrology   67
  surface water hydrology   71–73
  water cycle   68–69
  water quality hydrology   68, 75–76
hydropower   12, 368
hydropower, navigation, flood control
    interests   370
hypothetical reservoir   29
hypoxia/hypoxic waters   383

i

Imperial Irrigation District in California   12
Incident Command System (ICS)   324
income losses   4
industrial and urban development
    interests   368, 371
industrial wastewater   33
infiltration   85
Information Sharing and Analysis Center
    (ISAC)   329
information technology specialists   351
infrastructure bank   259, 284–285
inorganic contaminants   383
instream flow laws   307
instream flows   57, 182
intangible benefits   245
integrated model   124
integrated water management   383
Integrated Water Resources Management
    (IWRM)   2, 332
intensity-duration- frequency curves   85

interest rate formulas   249
International Council on Systems
    Engineering (INCOSE)   135
inventory and condition assessment   158
irrigation   12, 35–36, 52–54, 257
  losses   100
  systems   151
  water use   383
irrigators   251

j

Jerusalem   10
1889 Johnstown, Pennsylvania flood   312
justice system   289, 300–302

k

Kennedy, John F.   1
knowledge domains   119

l

landscape, irrigation, and water conservation
    specialists   351
land subsidence   14
large-scale dams   207–208
leaching   384
Lead and Copper Rule (LCR)   291
leak detection   154
legislation and regulatory controls   119
levee   384
lifecycle management   149
livestock water use   384
loan amortization   258–259
loan *vs.* bond comparison   259
local and regional water agencies   370
local ground water ordinances   297
London   10
low flow hydrology   73–75

m

maintenance management   154–155
management consultants   350
managers of financial, administrative, and
    legal affairs   352
man-made water resources system   384
Manning equation   108
master infrastructure development plans   346

maximum contaminant level (MCL) 297, 384
mechanical system 197
microbiology 10
microorganisms 384
1927 Mississippi River flood 312
Missouri River Master Manual 334, 337–339
mitigation 328–329
model and data, for decision making
    data types and sources 137–138
    hydrologic simulation model 142–146
    planning models and 146
    simulation models 138–141
    structure and functionality 136–137
modern turbines 12
MODFLOW 79
monetary returns 242
Multi-Criteria Decision Analysis
        (MCDA) 146, 198, 242, 258
multipurpose reservoirs 348

n

National Dam Inspection Act 317
National Dam Inspection Program
        (PL92-367) 317
National Environmental Policy Act
        (NEPA) 16, 235, 292, 298
National Flood Insurance Program
        (NFIP) 108, 313
National Incident Management System
        (NIMS) 324
National Inventory of Dams 27
National Pollutant Discharge Elimination
        System (NPDES) 294, 298
National Research Council 156
National Water Model 113
natural hydrologic elements 384
Natural Resources Conservation Service
        (NRCS) 17, 71, 179, 313
natural water infrastructure 38
NEPA background and process 238
nephelometric turbidity unit (NTU) 384
NOAA National Weather Service 112
noncommunity water system 385
Non-Community Water Systems 181
nongovernmental organizations (NGOs) 183

non-native species 385
nonpoint source controls 217
non-point source (NPS) pollution 385
non-resilient water systems 39
non-structural measures 385
non-structural problem-solving 2
non-transient, non-community water
        system 385
nonuniform flow 109
NPDES permit program 217

o

One Health 120, 121
operations management 150–152
organic contaminants 385
organic matter 385
osmosis 385
Ostrom, Elinor 245
outfall 385
oxygen demand 386

p

Palmer Drought Severity Index (PDSI) 322
Panama Canal 13
Paris 10
Partnership for Safe Water 160
Pecos River Compact 334, 335–336
percolation 386
performance indicators (PIs) 137
permeability 386
pesticide 386
pH 386
Philadelphia 10
photosynthesis 386
physical and life scientists 350–351
phytoplankton 386
pipeline repair 160
planners and policy specialists 352
planning
    basin 195
    environmental planning and assessment
        biodiversity problem 237–238
        environmental indicators and biological
            diversity 236–237
        guidelines for 238–239

NEPA background and process 238
tools for 239–240
infrastructure projects 347
model and data, for decision making 146
policy 194
for socio-political goals
principles and guidelines 226
social needs, water management
227–229
water management and social
welfare 225–226
water managers 229–231
for water infrastructure
conceptual to final plans 206–207
infrastructure lifecycle 204–205
large scale dams and river works
207–208
organizational types and sizes 205
problem-solving process 205
project development guidelines
205–206
urban water systems 208–209
water quality planning and management
assessment 218–219
CWA 220
enhanced data systems 219–220
human and natural determinants 216
management functions 214–216
program elements 216
science and management of 213–214
section 401 218
section 418 218
status of 212
water impairment 216
water resources planning 190
plant investment fee 257–258, 284
plumbing 15
point rainfall depths 83–84
point-source pollution 387
policy planning 194
polychlorinated biphenyls (PCBs) 387
polywrap 387
population growth and demand
forecasting 62
porosity 387
potable water 387

power and energy equations 42
power of turbine 42
precipitation 70, 104, 310
precipitation frequency data 84
preventive maintenance (PM) schedule 154
prior appropriation doctorine 387
private sector 10
problem identification 1, 193, 196, 197
programmable logic controllers (PLCs) 152
psychological concept 225
P-trap seal 11
public health 119
Public Health Act 15
public health engineering 15, 123
public information and education
specialists 351
public law 290
"Publicly Owned Treatment Works"
(POTWs) 33
public–private partnerships (PPP)
39, 245
public supply 48, 50, 387
Public Trust Doctrine 299
Public Utilities Commission (PUC) 294
public utility 181
Public Utility Regulatory Policies
Act 13
public water system (PWS) 388
public water use 388
pumping 34

q
quantitative demand 53

r
radionuclides 388
random component 144
rate increases, for wastewater utilities 347
rate of return (ROR) 253
rates and charges 292
rate-setting 286–287
rating curve 388
Rational Method 117–118
Rational Method formula 107
raw water 388

RCRA hazardous waste   388
reclaimed wastewater   388
recurrence interval   388
Red Fox Meadows Stormwater Project
      334, 340–341
regulatory organizations   367
reservoir   256–257
reservoir releases   29
reservoir storage-routing   109–112
reservoir water balance   101
return flow   389
revenue bond   389
reverse osmosis   389
revetment   389
Riparian Doctrine   296
riparian water rights   389
riprap   389
risk matrix   324
river basin   389
River Seine   10
river works   207–208
Rome   10
Rule of Law   290

S

sabotage   329
safe and reliable water   124–126
Safe Drinking Water Act (SDWA)   15, 19, 31,
      292, 294, 297, 306–307
safe yield   86
saline water   389
saltwater intrusion   14
San Francisco earthquake   321
sanitation   54
scientific and academic organizations   369
SDG goals   122
secondary drinking water standards   389
secondary wastewater treatment   390
security   229
sediment   390
sedimentation   28
self-supplied water   390
self-supply systems   46, 122
Senate Select Committee on Water
      Resources   17

septic tank   390
service access and quality   292
service providers   390
sewage treatment plant   390
shared governance   169–170
simulation models   135
small basin water balance   101
small water infrastructures   38
"smart" irrigation systems   12
smart water systems   153–154
social institutions   9
social justice   224
social welfare   224
social welfare function   248
socio-political goals
      principles and guidelines   226
      social needs, water management
            227–229
      water management and social
            welfare   225–226
      water managers   229–231
software vendors   155
Soil and Water Assessment Tool (SWAT)
      model   139
Soil Conservation Service (SCS)   313
sole source aquifer   390
South Florida Water Management District
      (SFWMD)   18
spillway   390
sprinkler   36
stakeholder   194, 390
stakeholder groups   370–372
standard project flood   391
state government   370
state regulatory agencies   19
storm sewer   391
stormwater   34–35, 309
stormwater flooding problems   309
stormwater management   55–56
stormwater rates   284
stormwater system   11
stormwater utility management   347
Strategic Environmental Assessment
      (SEA)   235
stream-aquifer   86

stream-aquifer interaction 297
streamflow 391
structural measures 391
subsidence 391
subsurface 36
Suez Canal 13
superfund 391
Supervisory Control and Data Acquisition
    (SCADA) system 135, 151
suppliers 372
suppliers and consultants 369
support sector 391
surface and groundwater data 137
surface water 391
surface water rights allocation 295
surface water hydrology 71–73
suspended sediment 391
suspended solids 391
sustainable development 11
Sustainable Development Goals (SDGs)
    120, 122, 242
system and process engineers 351
system failures 322

t

tangible benefits 245
tax exempt infrastructure bond 259
tax increment financing 391
tax revolts 11
technical sales and marketing 352
Tennessee Valley Authority (TVA) 12, 30,
    180, 246
tertiary wastewater treatment 392
thermal plants 12
thermal pollution 392
thermoelectric power water use 392
threatened and endangered species 392
total maximum daily loads (TMDLs)
    218, 298
transboundary conflicts 196
transboundary water issues 306
transboundary water law 299–300
transient, non-community water system 392
transmissibility 392

transmission pipes 31
transpiration 392
triangular hydrograph 107
tributary aquifer 78
triple bottom line (TBL) 197, 242
turbidity 392

u

underground storage tanks 392
underwriter 392
United Nations 242
United States 10, 33, 228
United States Code (USC) 291
United States water balance 48–50
units 43
universities, research institutes,
    publishers 371
unsaturated zone 393
Upper Colorado River Commission 181
urban demand 62
urban detention storage systems 310
urban farming 36
urban flood control 56
urban flooding 229
urban hydrology 11
urban systems 11
urban water audit 100–101
urban water demand 52
urban water services 393
USACE 177
USACE flood infrastructure 314
USACE's Hydrologic Engineering Center
    (HEC) 139
US Committee on Large Dams
    (USCOLD) 317
U.S. communities 11
US Constitution 190
US dam safety program 317
USEPA's Capacity, Management, Operations,
    and Maintenance (CMOM)
    program 161
US federal projects 198
US Geological Survey (USGS) 47
US–Mexico boundary 300

V

vandalism   329
variance   393
Virginia Beach water pipeline   336–337
vulnerability assessment   393

W

wastewater authority   4
wastewater charges   284
wastewater management   54–55
wastewater plant operation and
    recycling   347
wastewater treatment plant (WWTP)   294
water allocation law   292, 294–297
water balance
  irrigation water balances   94–95
  urban water balances   96
  water accounting in small watershed
    93–94
  water footprint and embedded water
    96–97
  in watersheds and river basins   90–92
  world, nations, states, and cities   92
waterborne disease outbreak   394
water-borne diseases   10
water charge calculation   284
water conflicts   289
water conservation   63
water contamination   123–124
water distribution system   51
Water Evaluation and Planning (WEAP)
    model system   146
water governance and institutions
  defined   165
  formal arrangements   166
  informal arrangements   167
  institutional arrangements   166
  policy and strategy   166
  regulatory controls   166
  scale factors in   168–169
  shared governance   169–170
  values and attitudes   167
water industry   173–175, 393
  employers   345–346

water infrastructure   2, 9
  broad issues of   38–39
  canals and pipelines for water
    conveyance   37–38
  components begin   25, 27
  dams, reservoirs, and hydropower
    systems   27–30
  demand for
    categories of water use   50–52
    direct requirements   46
    environmental water and instream
      flows   56–57
    flood control   56
    industrial and thermoelectric
      water uses   54
    irrigation   52–54
    sectors   46–47
    stormwater management   55–56
    United States water balance   48–50
    wastewater management   54–55
    water use data   47–48
  irrigation and drainage   35–36
  natural water infrastructure   38
  planning for
    conceptual to final plans   206–207
    infrastructure lifecycle   204–205
    large scale dams and river works
      207–208
    organizational types and
      sizes   205
    problem-solving process   205
    project development
      guidelines   205–206
    urban water systems   208–209
  river and coastal works   37
  small water infrastructures   38
  stormwater   34–35
  urban water systems   30
  variables   27
  wastewater systems   32–34
  water supply infrastructure systems
    31–32
water journals   373–376
water laws   18

allocation of water   290
instream flow laws   299
justice system and water litigation
    300–302
law and water management   293–294
public and private categories   290
regulatory and administrative
    law   290–292
Rule of Law   290
transboundary water law   299–300
types of water-related law   292–293
water litigation   289, 300–302
water management organizations
basin and regional water organizations and
    authorities   180–181
environmental law in   307
federal and state water agencies
    175–180
integration   183
regulators   182
support organization   182–183
water service providers   181–182
water management units   355–356
water managers   289, 310
water organizations   195
water pollution   120–122
water-power   12
water properties   358–359
water provider and wastewater agencies   367
water quality   122–123, 393
assessment   218–219
Clean Water Act   220
enhanced data systems   219–220
human and natural determinants   216
management functions   214–216
program elements   216
science and management of   213–214
section 401   218
section 418   218
status of   212
water impairment   216
water quality hydrology   75–76
water quality standards   393
water rates   283–284

water recycling systems   347
water-related public health   10
Water Resources Council   180, 219
Water Resources Development Act   313
water resources management
analytical problems   1
applications of   4
business of water   20–21
careers in   352–354
    asset management studies   347
    civil and environmental engineers   349
    construction and maintenance
        managers   350
    estuary recovery   348
    executives and managers of water
        agencies and utilities   349
    failures and remedial actions   348
    floodplain planning and
        management   348
    green infrastructure   347
    information technology specialists   351
    landscape, irrigation, and water
        conservation specialists   351
    management consultants   350
    managers of financial, administrative,
        and legal affairs   352
    master infrastructure development
        plans   346
    multipurpose reservoirs   348
    permitting   348
    physical and life scientists
        350–351
    planners and policy specialists   352
    planning infrastructure projects   347
    public information and education
        specialists   351
    rate increases, for wastewater
        utilities   347
    stormwater utility management   347
    system and process engineers   351
    technical sales and marketing   352
    wastewater plant operation and
        recycling   347
    water industry employers   345–346

water resources management (*cont'd*)
  water recycling systems 347
  water resources scientists are
    hydrologists and ecologists 349
  watershed manager 349–350
  water system operators 350
  common scenarios of 6
  dams 13–14
  defined 2–4
  economic of
    allocation 243
    assessment 245–246
    benefit–cost analysis 251–254
    defined 242
    natural and economic resources 243
    non-market values 243
    public and private economic
      goods 244–245
    time value of money 248–251
    water infrastructure investments
      246–248
  flood control 16
  flood risk/resolving conflicts 3
  food shortage 3
  government involvement in 16–18
  groundwater 14
  levels of responsibilities 2
  management structure 18–19
  management variables of 6
  monitoring and assessment 3
  public health engineering 15
  purposes of 4–7
  quality and environmental
    protection 15–16
  science, engineering, and technology 19
  scientific and management knowledge 1
  types of responsibilities 2
  water law 18
water resources planning 190

Water Resources Planning Act 189, 191,
  219, 226
water resources scientists are hydrologists and
  ecologists 349
water reuse systems 11
water right owner 295
water rights 99–100, 256
water rights law 306
water rights system 296
water scarcity 339
water security
  drought and climate shifts 322–323
  elements 319–320
  risk management
    and disaster preparedness 323–326
    mitigation 328–329
    terminology of 326–327
  threats and impacts 321–322
  vulnerability assessment 328
watershed 47, 394
Watershed Academy 220
watershed manager 349–350
water supply 9
water system 159–162
  applications to 159–162
  operators 350
  smart water systems 153–154
water table 393
water use data 47–48
water use sectors 9–13
water utilities 10
wellhead protection area 394
wetland ecosystems 394
World Bank 205, 242
World Commission on Dams 208
World Water Council 331

X
xeriscape 394